# Engineering Applications of Computational Methods

Volume 24

**Series Editors**

Liang Gao, State Key Laboratory of Digital Manufacturing Equipment and Technology, Huazhong University of Science and Technology, Wuhan, China

Akhil Garg, School of Mechanical Science and Engineering, Huazhong University of Science and Technology, Wuhan, China

The book series Engineering Applications of Computational Methods addresses the numerous applications of mathematical theory and latest computational or numerical methods in various fields of engineering. It emphasizes the practical application of these methods, with possible aspects in programming. New and developing computational methods using big data, machine learning and AI are discussed in this book series, and could be applied to engineering fields, such as manufacturing, industrial engineering, control engineering, civil engineering, energy engineering and material engineering.

The book series Engineering Applications of Computational Methods aims to introduce important computational methods adopted in different engineering projects to researchers and engineers. The individual book volumes in the series are thematic. The goal of each volume is to give readers a comprehensive overview of how the computational methods in a certain engineering area can be used. As a collection, the series provides valuable resources to a wide audience in academia, the engineering research community, industry and anyone else who are looking to expand their knowledge of computational methods.

This book series is indexed in both the **Scopus** and **Compendex** databases.

Xu Cheng · Fan Shi · Xiufeng Liu · Shengyong Chen

# Computational Methods for Blade Icing Detection of Wind Turbines

Springer

Xu Cheng
Tianjin University of Technology
Tianjin, China

Fan Shi
Tianjin University of Technology
Tianjin, China

Xiufeng Liu
Technical University of Denmark
Lyngby, Denmark

Shengyong Chen
Tianjin University of Technology
Tianjin, China

ISSN 2662-3366 ISSN 2662-3374 (electronic)
Engineering Applications of Computational Methods
ISBN 978-981-96-6762-8 ISBN 978-981-96-6763-5 (eBook)
https://doi.org/10.1007/978-981-96-6763-5



This Springer imprint is published by the registered company Springer Nature Singapore Pte Ltd.
The registered company address is: 152 Beach Road, #21-01/04 Gateway East, Singapore 189721, Singapore

# Preface

This book is designed to encompass several cutting-edge computational methodologies applied to the critical challenge of blade icing detection in wind turbines. The increasing reliance on wind energy as a sustainable power source necessitates robust and efficient solutions to mitigate the detrimental effects of blade icing. The chapters herein delve into advanced computational techniques for accurately and reliably detecting blade icing, addressing a problem of significant practical and economic importance. There are eleven chapters in this book, and they are organized as follows.

Part I, *Temporal Convolutional Neural Network-Based Methods*, lays the foundation by exploring deep learning techniques centered around Temporal Convolutional Neural Networks (TCNs). This part emphasizes the potential of advanced TCN architectures to deliver robust, precise, and practically viable solutions for blade icing detection.

- Chapter 2, "Attention-Enhanced Temporal Convolutional Neural Network for Blade Icing Detection," introduces a novel Temporal Attention Convolutional Neural Network (TACNN) that incorporates a temporal attention mechanism to discern the significance of individual sensors and time steps, thereby enhancing feature extraction and capturing crucial temporal dependencies in time-series data.
- Chapter 3, "Wavelet-Enhanced Temporal Convolutional Neural Network for Blade Icing Detection," develops an innovative TCN architecture, the Wavelet-Enhanced Temporal Convolutional Neural Network, engineered to capture comprehensive features across both time and frequency domains through the integration of Multilevel Discrete Wavelet Decomposition and a Multilevel Convolutional Recurrent Neural Network.
- Chapter 4, "Class-Imbalanced Semi-supervised Temporal Neural Network for Blade Icing Detection," tackles the limitations of traditional supervised learning in the context of imbalanced blade icing datasets by proposing a novel Deep Class-Imbalanced Semi-Supervised (DCISS) model, a prototypical network designed to learn from both labeled and unlabeled data.

Part II, *Graph Convolutional Neural Network-Based Methods*, shifts the focus to graph neural network-based approaches, which leverage the inherent connectivity and relationships between various components of a wind turbine system. This part highlights the potential of GNN methodologies to enhance modeling complex relationships and extract meaningful information for blade icing detection.

- Chapter 5, "Attention-Enhanced Graph Convolutional Neural Network for Blade Icing Detection," introduces the Attention-Enhanced Graph Convolutional Neural Network (GTAN), a novel GNN architecture utilizing graph convolutional networks and temporal attention mechanisms to explicitly model sensor relationships and capture spatial-temporal dependencies for improved icing detection.
- Chapter 6, "Wavelet-Empowered Graph Convolutional Neural Network for Blade Icing Detection," presents a Wavelet-Empowered Graph Convolutional Neural Network (MWGCN) that combines multiscale wavelet decomposition with graph neural networks, using a scale attention module to enhance feature learning across different scales and effectively harness temporal data.
- Chapter 7, "Class-Imbalanced Graph-Temporal Convolutional Neural Network for Blade Icing Detection," addresses the challenge of class imbalance in graph-based methods by introducing a Class-Imbalanced Graph-Temporal Convolutional Neural Network, featuring a distance-based classifier specifically designed to mitigate bias towards the majority class in imbalanced datasets.

Part III, *Federated Convolutional Neural Network-Based Methods*, explores federated learning (FL) methodologies, which enable collaborative model training among geographically dispersed wind farms without directly sharing sensitive raw data. This part emphasizes the potential of federated learning to facilitate collaborative model training, protect data privacy, and improve model performance through novel approaches for knowledge transfer and class imbalance handling.

- Chapter 8, "Split Heterogeneous Federated Convolutional Neural Network for Blade Icing Detection," delves into the complexities of federated learning and introduces a novel heterogeneous federated learning approach (BiFL) designed to function efficiently across different wind turbine types while addressing data imbalance, employing split learning and knowledge distillation.
- Chapter 9, "Data-Efficient Heterogeneous Federated Convolutional Neural Network for Blade Icing Detection," presents FedBIP, a Data-Efficient Heterogeneous Federated Convolutional Neural Network that leverages feature selection and knowledge distillation within a federated learning framework for enhancing training and prediction performance in data-scarce settings.
- Chapter 10, "Blockchain-Empowered Clustered Federated Convolutional Neural Network for Blade Icing Detection," explores the integration of blockchain technology into a clustered federated learning framework, presenting a Blockchain-Empowered Clustered Federated Learning Model (BLADE) for creating more secure and robust federated learning for blade icing estimation.

This book provides a comprehensive overview of advanced computational methods for blade icing detection, offering both theoretical insights and practical implementations. The methodologies detailed herein demonstrate the potential of deep learning, graph neural networks, and federated learning to significantly enhance the accuracy, robustness, and efficiency of blade icing detection systems. It is our hope that this book serves as a valuable resource for researchers, engineers, and practitioners working to advance the field of wind energy and improve the reliability and sustainability of wind turbine operations in cold climates.

Tianjin, China Xu Cheng
Tianjin, China Fan Shi
Lyngby, Denmark Xiufeng Liu
Tianjin, China Shengyong Chen

**Acknowledgements** This work was partially supported by the National Natural Science Foundation of China under Grants 62306212, T2422015, and 62020106004, as well as by the Tianjin Natural Science Foundation under Grant 23JCJQJC000070. Xu Cheng also received support from the Marie Skłodowska-Curie Actions (MSCA) under Project No. 101111188.

**Competing Interests** The authors have no competing interests to declare that are relevant to the content of this manuscript.

# Contents

## Part II Graph Convolutional Neural Network-Based Methods

# Chapter 1
# Introduction

## 1.1 Background

With the continuous development of human society and economy, global energy consumption is projected to double by 2050 [1]. Simultaneously, humanity faces the goal of achieving carbon neutrality by mid-century. In the future energy structure, we face a greater demand for energy consumption coupled with stricter carbon emission standards. The key to resolving this contradiction undoubtedly lies in renewable energy sources. Among these, wind energy, currently the most commercially utilized and successful renewable energy, undoubtedly bears an even greater mission. Wind energy is widely distributed, abundant in reserves, relatively inexpensive to develop, environmentally friendly, rapidly producible, and renewable, providing strong support for future energy development. According to statistics from the U.S. Energy Information Administration (EIA) and the American Wind Energy Association, wind energy's share of U.S. energy consumption surpassed hydropower in 2019, becoming the largest source of renewable energy consumption in the United States [2]. In the first nine months of 2020, wind energy generated 2.156 quadrillion British thermal units (Btu), while hydropower generated 2.031 quadrillion Btu. A mid-2020 report by the climate and energy think tank Ember indicated that renewable energy generation in EU countries exceeded fossil fuel generation, accounting for 40% of electricity consumption across the 27 EU member states, while fossil fuels accounted for only about 34%. Within renewable energy, wind and solar power accounted for 12% of new energy generation, setting a new historical record. In certain new energy technology powerhouses such as Denmark, Ireland, and Germany, wind and solar power accounted for as high as 64, 49, and 42% of electricity generation respectively [3]. China, with its vast territory, possesses exceptionally abundant wind energy resources for development and utilization. China established its first demonstrative wind power plant in Dabancheng, Xinjiang in 1989. From 1989 to the present, China has maintained a global leading position in both new installed capacity and total installed capacity for many consecutive years. It can be said that China's

 
X. Cheng et al., *Computational Methods for Blade Icing Detection of Wind Turbines*,
Engineering Applications of Computational Methods 24,
https://doi.org/10.1007/978-981-96-6763-5_1

wind power industry is transitioning from an alternative energy source to a primary energy source. Statistics from the China National Energy Administration show that in 2019, China's cumulative installed wind power capacity reached 210.05 million kilowatts, accounting for approximately 32.29% of the global cumulative installed wind power capacity. In the first half of 2020, China's newly grid-connected wind power capacity reached 6.32 million kilowatts [4]. In the foreseeable future of energy development, wind energy will continue to maintain strong momentum, and it will inevitably maintain a leading position in energy consumption.

With the continuous advancement of wind power technology, wind turbines are progressively evolving from small and medium-sized to large and ultra-large scale [5–8]. Regarding current wind power technology, due to limitations in wind farm site selection, turbine transportation, and installation, onshore wind farms predominantly utilize wind turbines with a single unit capacity of around 2 MW. However, for offshore wind farms, the majority of wind turbines are 5 MW or larger units. Literature indicates that 8 MW output wind turbines have already been deployed in practical applications [9]. Wind turbine designs in the 10–20 MW power range are in the research and development phase [10]. Furthermore, the U.S. Department of Energy (DOE) has initiated and funded research and development efforts for 50 MW output wind turbines at relevant U.S. universities and research institutions. The trend towards larger wind turbines effectively enhances wind energy capture efficiency and significantly reduces the cost of wind power generation, thereby powerfully promoting the development of the entire wind power industry. Concurrently, with the rapid development of wind energy, its proportion in the national economy will continue to increase. This places higher demands on the reliable and stable operation of wind turbine units in complex natural environments.

According to relevant theories, from sea level up to an altitude of 1000 meters, wind speed increases by approximately 0.1 meters per second for every 100 meters of altitude gain [11]. Given that wind energy is proportional to the cube of wind speed, the increase in altitude significantly enhances wind energy utilization. Consequently, to maximize wind energy capture, wind farms are often constructed on ridgelines, mountaintops, or other high-altitude locations. Statistics indicate that approximately one-third of the world's wind farms are built in high-altitude or cold regions such as Northern Europe, China, and Canada [12]. While wind farms built in these regions benefit from increased power generation, they also endure the impacts of extreme climates associated with high-altitude and cold regions. Among these impacts, wind turbine blade icing is a primary concern. To address the consequences of wind turbine blade icing, the International Energy Agency (IEA) has established a dedicated working group on wind energy in cold climates. This group has summarized and analyzed wind power technologies in cold regions and made corresponding arrangements for future research efforts [12–14]. In the early stages of wind power development in China, the exploitation of wind resources was mainly concentrated in the northwestern regions. These regions have dry climates, and the problems caused by wind turbine blade icing were not significant. However, with the rapid development of wind farms in southern China in recent years, wind turbine blade icing has gradually gained attention. This is primarily because the southern climate is humid, with high altitudes and low temperatures, leading to frequent icing in some southern wind farms during winter. In wind-resource-rich areas of southern China, wind turbine blade icing has progressively become a major factor affecting the normal operation and development of wind farms [15].

Starting in the late 1990s, international researchers have conducted extensive research on the impact of wind turbine blade icing on turbine performance [11, 16–18]. According to relevant studies, under heavy or severe icing conditions, wind turbine blade icing can lead to a reduction in annual power generation of approximately 10 to 20%. Other studies on wind turbine blade icing have also indicated that the proportion of annual power generation loss due to blade icing ranges from 5 to 25% [19, 20]. A Spanish research group analyzed 29 months of operating data from 517 wind turbines in Spain under icing conditions. Their extrapolation estimated that annual power generation losses due to blade icing across the country were as high as 550 GWh, equivalent to the electricity consumption of 200,000 households [21]. Research tracking wind turbines in Canada and Sweden showed that blade icing resulted in annual power generation losses of 6.6% and 16.0%, respectively [22].

With the development of wind power in China, accidents caused by wind turbine blade icing have been frequently reported. In early 2008, over 20 days of freezing weather caused power transmission towers to collapse and power lines to break in Guizhou Province's power grid. In 2015, a wind farm in Guizhou with 48 wind turbines and a total output power of 96 MW was affected by intermittent icing weather for six months, resulting in a total downtime of 42 days and a power generation loss of 72,700 kWh. According to statistics, annual power generation losses due to winter freezing weather in the Guizhou region are as high as 15 to 20%. The Pingshan Wind Farm in Yudu, Jiangxi, commissioned in 2014 at an altitude of 1300 meters, also experienced several wind turbine blade icing events in January 2015 due to small-scale icing weather, resulting in a power generation loss of approximately 39,000 kWh. The Yangmeishan Wind Farm in Luliang, Yunnan, and the Subaoding Wind Farm in Hunan have both experienced wind turbine blade icing accidents leading to shutdowns due to icing weather. Research on wind turbine blade icing has gained attention with the rise of wind power generation. Table 1.1 provides a preliminary summary of the impact of wind turbine blade icing on turbine performance and power generation.

The impact of wind turbine icing is multifaceted. First, turbine operation heavily relies on measurement instruments on the nacelle, such as anemometers and wind vanes. Icing poses a significant challenge to the measurement accuracy of these instruments. When these instruments fail to function properly due to icing, erroneous pitch and yaw control commands may be issued, leading to reduced power generation at best and mechanical structural damage at worst. Second, wind turbine blade icing alters the blade airfoil shape, increasing drag and reducing lift, thereby decreasing power generation. Third, wind turbine blade icing introduces new structural issues. During icing and de-icing processes, the ice distribution on the blades is constantly changing [32]. Blade icing reduces the blade's natural frequency. The trend of ice growth and change can be observed in references [33, 34]. Fourth, icing causes

**Table 1.1** Summary of research on power generation losses caused by wind turbine blade icing

| No. | Research content | Research findings |
|---|---|---|
| 1 [23] | Evaluated the impact of wind turbine blade icing on power generation by combining observations from actual wind turbine tests and computer simulations. | Without de-icing operations, power generation losses caused by wind turbine blade icing accounted for approximately **17%** of annual power generation. With de-icing operations, losses decreased to 2% |
| 2 [16] | Based on wind turbine blade calculation models and actual climate measurements, assuming turbines are stopped when icing occurs. | Without de-icing operations, power generation losses caused by wind turbine blade icing accounted for approximately 10% of annual power generation. With de-icing operations, losses decreased to **3.5%** |
| 3 [24] | Used commercial wind turbine operating data from Sweden | Power generation losses due to blade icing accounted for approximately 6.4% of annual power generation. Turbine downtime due to icing accounted for **2.1%** of total operating time |
| 4 [25] | Operating data from 23 wind farms in Canada. | Power generation losses due to blade icing in Canada accounted for approximately **3.9%** of annual power generation, varying by province due to different severity levels |
| 5 [26] | Blade icing experiments conducted in a wind tunnel | Icing resulted in approximately 16–26% power generation loss |
| 6 [27] | Simulated blade icing using LEWICE, performed aerodynamic analysis using FLUENT, and generated power curves using BEMT | Power generation loss was 35% below rated wind speed. Power at cut-in wind speed was 50% lower than in non-icing conditions, and power decreased with increasing wind speed |
| 7 [28] | Simulated 5 MW virtual wind turbine blade icing using FENSAP-ICE and estimated power curves using BEMT code. | Power generation loss between cut-in and rated wind speed was 28% |
| 8 [29] | Wind tunnel simulation of wind turbine blade icing | **62%** power generation loss |
| 9 [30] | Wind turbine data from the Gaspé Peninsula, Canada | For 27 actual icing events, instantaneous power generation losses ranged from 7–96% |
| 10 [31] | Proposed a numerical model applicable to wind turbines based on helicopter rotor icing models | Up to 40% power generation loss |

aerodynamic performance imbalance, increasing the load on mechanical structures such as towers, bearings, and gearboxes, increasing the risk of failure and shortening their service life. Icing also leads to an increase in nacelle vibration acceleration [35]. According to relevant research, wind turbine lateral motion is caused by mass and aerodynamic imbalance, while downward motion is caused solely by aerodynamic imbalance. Increased vibration also shortens the fatigue life of towers and other structural components. Changes in airfoil aerodynamic characteristics due to icing cause greater damage to fatigue life than changes in ice mass [36]. The effect of low temperature on oil viscosity and component dimensional changes may also lead to failures of mechanical components (such as bearings and gearboxes). Fifth, wind turbine blade icing also leads to issues such as ice throw. During icing, a large amount of ice accumulates near the blade tip. When ice is thrown from this location, it has a higher initial velocity and is thrown to a greater distance, posing a significant risk to residents near wind farms. Furthermore, wind farms often need to shut down wind turbines once ice throw occurs, leading to considerable economic losses. Sixth, wind turbine blade icing also introduces noise problems. Wind turbine blades generate noise due to the interaction between the turbulent boundary layer and the airfoil trailing edge, periodic vortex shedding, and airflow separation during stall [37–39]. More seriously, even a small amount of icing can cause higher noise levels [37]. Considering the health of the public surrounding wind farms, this noise is also a factor contributing to wind farm shutdowns.

Through the above analysis, it is evident that if icing faults on wind turbine blades are not detected and eliminated in a timely manner, it can, at best, exacerbate the wear and tear of wind turbines, and at worst, lead to blade fracture, causing accidents with casualties and equipment damage. In light of this, to effectively reduce turbine maintenance costs and operational risks, and to improve the safety and economic viability of wind turbines, it is highly necessary to vigorously develop wind turbine blade icing detection technology. This is of theoretical and practical value for enhancing the operational reliability of wind turbines and reducing icing hazards.

## 1.2 State of the Art

### *1.2.1 Icing Factors on Wind Turbine Blades*

Wind turbine blade icing refers to the phenomenon of supercooled water droplets freezing on the surface of wind turbine blades during operation [40, 41]. Supercooled water droplets are liquid water with temperatures below 0 °C but not yet frozen. According to relevant research, the temperature of supercooled water droplets can be as low as minus 30 °C without freezing[42, 43]. Icing on wind turbine blades has varying sizes, shapes, and characteristics. Literature typically defines it based on factors such as Mean Volume Diameter (MVD) of droplets, Liquid Water Content (LWC), air temperature, wind speed, and blade length [44, 45]. The appearance of icing on wind turbine blades is influenced by the ambient temperature, with distinct appearances at different temperatures [46].

There are three possible factors for wind turbine blade icing: Atmospheric in-cloud icing [27], Precipitation Icing, and Frost Icing. Atmospheric icing can occur when the average droplet size and liquid water content of supercooled water droplets in the air are small, and the temperature is below minus 30 °C, causing them to impact and freeze on wind turbine blades. Ice formed in this case is usually rough, opaque, low in density, and has weak adhesion. The temperature and size of supercooled water droplets also determine whether the formed ice is rime ice, hoar frost, or glaze ice [15]. Rime ice forms when larger supercooled water droplets spread out on the blade surface upon impact, forming a water film that then freezes [47]. Hoar frost is formed when smaller supercooled water droplets spread out on the blade surface upon impact, forming a water film that then freezes. Glaze ice forms when supercooled water droplets do not immediately freeze upon impact with the blade surface, remaining mostly in liquid form before freezing. Glaze ice formation is mainly due to fewer air pockets trapped within the ice layer. Precipitation icing occurs in the form of freezing rain or wet snow, which freezes on the blade surface upon impact. Freezing rain occurs when rain falls onto a surface with a temperature below 0 °C. This phenomenon is mainly caused by temperature inversion, where cold air is trapped near the ground beneath a layer of warmer air. Temperature inversion can also occur above wind turbine blades when the blade temperature is significantly below freezing and the air temperature is above freezing. Ice formed from freezing rain is typically transparent, with an amorphous and dense structure, and it adheres firmly to the blade surface with very strong adhesion. Wet snow icing conditions exist when slightly liquid snow (wet snow) at air temperatures between 0 °C and 3 °C impacts and adheres to the blade surface. Due to the presence of some liquid water in wet snow, snow crystals and liquid water combine when snow crystals come into contact with the blade surface. If the subsequent temperature drops below 0 °C, very strong and difficult-to-remove icing will form on the wind turbine blade surface. Frost icing occurs when the blade surface temperature is below the air dew point. In this case, water vapor in the air will deposit on the wind turbine blade surface, forming small ice crystals.

Wind turbine blade icing can generally be divided into stages such as early icing, noticeable icing, severe icing, and natural recovery (de-icing). Early icing refers to the initial appearance of icing on the blade surface when ambient temperature and other icing conditions are met. Due to the delay in icing development, early icing on wind turbine blades is not obvious and has little impact on the unit. If early icing is not detected, it will evolve into noticeable and severe icing stages. In these stages, the operating efficiency of the wind turbine is affected, leading to reduced power generation and unstable wind turbine operation. After wind turbine blade icing persists for a period, it will melt or shed (or possibly due to artificial de-icing). This stage is called the natural recovery stage. Without human intervention, the natural recovery stage will take longer than the previous stages.

Factors affecting wind turbine blade icing can be summarized in the table (Table 1.2):

**Table 1.2** Factors influencing wind turbine blade icing

| No. | Factor | Mechanism of icing influence |
|---|---|---|
| 1 | Temperature | Temperature is the main factor affecting the heat transfer characteristics of water droplets, influencing the icing coefficient of water droplets. The higher the temperature, the smaller the icing coefficient of water droplets. Conversely, the lower the temperature, the larger the icing coefficient of water droplets. When the icing coefficient is small, the icing type on the blade is mainly clear ice; when the icing coefficient is large, the icing type on the blade is frost ice |
| 2 | Liquid water content | Liquid water content affects the icing thickness. The higher the liquid water content, the thicker the icing on the blade; conversely, the lower the liquid water content, the thinner the icing on the blade |
| 3 | Droplet diameter | Droplet diameter has a similar effect on blade icing as liquid water content. When the droplet diameter is large, it also has different degrees of influence on the freezing and adhesion characteristics of droplets, droplet impact, and icing process. The larger the droplet diameter, the thicker the icing; conversely, the thinner the icing |
| 4 | Wind speed | Wind speed affects the impact velocity of water droplets on the blade. The greater the wind speed, the faster the convective heat transfer on the blade surface, and the greater the amount of icing; the smaller the wind speed, the slower the convective heat transfer on the blade surface, and the smaller the amount of icing |
| 5 | Airfoil | The airfoil of the blade determines the aerodynamic load characteristics of the wind turbine, thereby affecting the icing condition of the wind turbine blade |
| 6 | Surface roughness | Blade surface roughness mainly affects the aerodynamic characteristics of the blade airfoil. The increase of blade surface roughness increases the friction between the blade and the air, resulting in a decrease in the lift coefficient |

Jiang Wenhua et al. studied the effect of blade surface roughness on the wind energy utilization factor of wind turbines. The research results found that the aerodynamic characteristics of wind turbine blades increase with the increase of the angle of attack, but they change due to the change in surface roughness caused by blade icing [48]. Caccia et al. concluded through numerical simulations of wind turbine blade icing that the icing position on the blades is concentrated at the leading edge, and the position with the most icing is concentrated at the stagnation point [49]. Manatbayev et al. analyzed the causes and required conditions, influencing factors, etc., of wind turbine blade icing by selecting the prediction model and thermodynamic model required for blade icing [50]. Ibrahim et al. conducted numerical simulations on the static icing of vertical-axis wind turbine blades and obtained information such as the ice type of icing by comparing icing under different airfoil shapes [40].

Currently, many domestic and foreign research institutions and enterprises have conducted in-depth research on wind turbine blade icing detection/de-icing technologies. According to different usage methods, wind turbine blade icing detection methods are mainly divided into three categories: traditional anti-icing/de-icing methods, mechanism model-based methods, and data-driven methods.

## 1.2.2 Traditional Anti-Icing/De-Icing Methods

Ice accretion on wind turbine blades shares numerous similarities with ice formation on aircraft wings. Consequently, traditional research methodologies for wind turbine blade icing have largely drawn inspiration from the aviation industry. However, this approach has often overlooked the unique characteristics inherent to wind turbine blade icing [51]. Firstly, in terms of blade geometry, thermal properties, operational conditions, and icing conditions, there are notable distinctions between wind turbine blade icing and aircraft wing icing. Wind turbine blades are composed of a series of specialized airfoils that undergo significant geometric variations from the blade tip to the root. Furthermore, each blade within a turbine exhibits distinct twist angles. In contrast, aircraft wings typically employ symmetrical and thin airfoil designs. Considering both aerodynamic and structural requirements, specialized airfoils for wind turbines are generally asymmetrical and possess greater thickness compared to aircraft wings. Secondly, the flow field surrounding wind turbine blades is fundamentally different from that around aircraft wings. Due to the rotation and pitch adjustments of wind turbine blades during operation, the resultant velocity along both the spanwise and chordwise directions exhibits substantial variations. This, in turn, profoundly affects the water droplet impingement process, water runoff behavior, and convective heat transfer during icing. Moreover, the thermal conductivity of wind turbine blades, often constructed from polymer composites, is considerably lower than that of metal-based aircraft wings. This disparity significantly influences the icing process and the selection of anti-icing/de-icing strategies. Thirdly, wind turbine icing predominantly occurs during cold winter weather, which contrasts with aircraft icing, typically encountered at higher altitudes when traversing clouds. When aircraft traverse clouds in the stratosphere, wing icing often occurs, with the icing process potentially lasting from seconds to minutes. However, wind turbines situated in the troposphere are subjected to more complex meteorological conditions. Beyond in-cloud icing (i.e., freezing fog), phenomena such as freezing drizzle/sleet, wet snow, and frost can also induce severe ice accretion on wind turbine blade surfaces, potentially persisting for hours or even days. Furthermore, wind turbines are more likely to be exposed to environments with high liquid water content (LWC) and larger droplet sizes, particularly in mountainous and offshore regions, leading to more substantial ice accumulation. Concurrently, offshore wind turbines are even susceptible to icing induced by salt spray/sea spray. Fourthly, aircraft wing icing is typically an isolated event occurring on individual aircraft, which can be considered an isolated/random event. In contrast, wind farms comprise dozens or even hundreds of turbines, rendering wind turbine blade icing a widespread event. This makes ice accretion on wind turbine blade surfaces more predictable on a broader scale.

Current anti-icing/de-icing technologies commonly employed in wind farms can be summarized as follows:

**Heating Methods**. Heating methods are highly efficient and technologically mature, representing the most prevalent approach for mitigating wind turbine blade icing. Heating methods can be further categorized into hot air injection, electrothermal methods, and other thermal methods. Hot air injection utilizes the internal circulation of heated air within the turbine to prevent or remove ice accumulation on blade surfaces. Companies such as Enercon, Vestas, and GE Wind have commercialized this technology for de-icing systems [52]. Electrothermal methods, or resistive heating, function by directly elevating the blade surface temperature to prevent ice formation. Siemens, Vestas, Nordex, and WinWind [14] employ this technology for anti-icing/de-icing on the outer 1/3 of the full blade chord and the outer 2/3 [53] of the leading edge. Other thermal methods encompass diverse forms of energy, including microwave heating [54], plasma heating [55], and infrared heating. While thermal methods have achieved widespread application, they are not without limitations:

- Firstly, hot air-based thermal methods rely on heat transfer to warm the blade surface. However, the low thermal conductivity of polymer composite wind turbine blades, especially large-scale blades based on epoxy resins, significantly increases the energy demand for anti-icing/de-icing processes. This approach also necessitates additional fan heaters, making it challenging to retrofit existing operational wind turbines.
- Secondly, resistive heating methods can be installed externally or internally on the blades. Compared to hot air methods, they offer greater effectiveness, flexibility, and ease of retrofitting for existing turbines [56]. However, external installation on blade surfaces may increase surface roughness, potentially impacting aerodynamic characteristics [57]. According to relevant literature, when employing resistive heating for wind turbine blade anti-icing/de-icing, focusing heating on the blade leading edge rather than the entire blade is recommended to conserve energy [54, 57, 58].
- Thirdly, heating methods such as microwave heating, due to the potentially high voltage requirements of their energy sources, may introduce electromagnetic interference to other turbine equipment and elevate the risk of lightning strikes.

**Anti-icing Coatings**. Anti-icing coatings utilize their specific operational characteristics to reduce the amount of ice accumulating on blade surfaces [59–63]. These coatings can be readily applied to wind turbine blade surfaces. Although coatings may not completely eliminate ice formation, their low installation and near-zero operational costs render them highly attractive to researchers and wind farm operators in recent years. Due to their high relevance to practical applications, field tests are crucial to validate the performance and reliability of various coatings. Successful field testing necessitates close collaboration among research institutions, turbine manufacturers, and wind farm operators, alongside supportive government policies and substantial financial investment. To date, publicly available information on field tests of anti-icing coatings remains limited.

**Control-based Methods**. Control-based methods aim to decelerate ice accumulation by adjusting blade pitch angles and rotational speeds [57]. Adjusting wind turbine blades can regulate the pitch angle and angle of attack, which, in turn, can reduce ice adhesion on blade surfaces, thereby mitigating aerodynamic losses [64]. During severe icing events, operating only wind turbines equipped with anti-icing systems may be considered. When even anti-icing systems become ineffective, proactively stopping all wind turbine operations is advisable to minimize ice accumulation on blades. This also facilitates faster ice melting, enabling turbines to resume operation sooner [57]. Furthermore, a slight reduction in rotor speed can lessen ice accretion on blade surfaces. Control-based methods offer advantages such as high safety, low energy consumption, and long-term benefits. However, the relationship between wind turbine operational status and the impact of blade surface icing on turbine aerodynamic performance is not yet fully understood, necessitating further research.

**Mechanical Methods**. Drawing from aviation practices, mechanical methods employ various vibrations [65, 66] or movements [67] to dislodge ice accumulated on blade surfaces. Habibi et al. developed an ultrasonic de-icing system utilizing piezoelectric actuators to detach ice layers from surfaces [68]. While ultrasonic de-icing is effective and energy-efficient, the requirement for ultrasonic transducers to be installed on the inner surface of turbine blades reduces the flexibility of this technology for application in existing turbines. Other mechanical methods used in the aviation industry, such as pneumatic de-icing boots that inflate and deflate to deform the ice layer for removal, are difficult to install on wind turbine blades and pose a risk of degrading blade aerodynamic characteristics. Moreover, when mechanical de-icing techniques are used while blades are rotating in anti-icing mode, detached ice fragments may present safety hazards to nearby turbines, vehicles, or pedestrians.

Literature [11] critically reviewed the advantages and disadvantages of various anti-icing/de-icing technologies developed for wind turbines prior to 2009. Fakorede et al. analyzed and compared various anti-icing/de-icing technologies developed before 2015 from a market perspective. Currently, hybrid strategies combining two or three anti-icing/de-icing technologies offer a potential solution for achieving enhanced de-icing performance. For currently operational turbines, combining external resistance heaters, coatings, and control-based methods may represent the only viable option. However, for new turbine designs, a broader range of choices exists, such as utilizing hot air injection or ultrasonic de-icing. Future research is warranted to further develop hybrid anti-icing/de-icing strategies for wind turbines.

### 1.2.3 Mechanism-Based Model Methods

Mechanism-based models primarily rely on constructing mathematical models based on the internal mechanisms of the subject under study. In this research context, physical models are chiefly developed by leveraging the mechanisms of wind turbine blade icing to determine the presence of icing on wind turbine blades. Lasse Makkonen et al. proposed a numerical model to simulate ice accretion and ice shape on wind turbine blades. This model simulated both rime ice caused by fog droplets and glaze ice caused by precipitation across all droplet attack angles on wind turbine blades. It also simulated icing conditions under blade heating and investigated the sensitivity of blade icing to meteorological factors. The model's usability and superiority were validated by comparing its predictions with data from wind tunnel experiments and field studies of natural wind turbine blade icing [69]. Saleh et al., through constructing a multi-scale model based on wavelet analysis for wind turbines, discovered that blade icing increases inertia. They also found that changes in generator current frequency could be analyzed to predict wind turbine blade icing [70]. Davis et al. presented three methods for analyzing wind turbine power curves to predict blade icing and validated the superiority of these methods using real operational data from four wind farms [71]. Wang et al. introduced a numerical simulation model capable of analyzing wind turbine blade icing in three dimensions [72]. The effectiveness of this model was confirmed by comparing predicted icing results with actual icing scenarios. Hu Qin et al. conducted numerical simulations and actual blade icing experiments on NE-100 wind turbine blades, obtaining relevant models and icing morphologies for blade surface icing under different external environments [73]. Kabardin et al. designed an ice detection device based on piezoelectric ceramics for wind turbine blade icing. Based on the mechanism of blade icing, this device can effectively measure the thickness of blade ice accretion [74]. Ping Fu et al. simulated the rime ice accretion process on horizontal axis wind turbines using the commercial CFD software Fluent [75]. This model analyzed the impact of meteorological factors such as wind speed, mean volume diameter of water droplets, liquid water content, and air temperature on wind turbine blade icing. Furthermore, the model can calculate the three-dimensional shape of ice accretion. Homola et al. analyzed the influence of air temperature and droplet size on wind turbine blade icing. Numerical simulation studies on a 5MW pitch-controlled wind turbine revealed that ice growth increases with larger droplet sizes and that changes in atmospheric temperature significantly affect the shape of ice accretion. Streamlined ice was observed at low temperatures, while horn-shaped ice was observed at higher temperatures [76, 77]. Deng Xiaohu et al. simulated the airflow field before and after wind turbine blade icing, determined the shape of ice accretion, and discussed the impact of wind turbine blade icing on aerodynamic characteristics such as blade lift and air resistance. Manatbayev et al. proposed an airflow field model capable of analyzing wind turbine blade icing in three dimensions, studying wind turbine blade icing from the perspective of wind turbine aerodynamic characteristics [50]. Sierikova et al. investigated wind turbine blade icing using turbulence models, finding that clear ice attached to the blades has a greater impact on wind turbine aerodynamic characteristics than frost ice [78].

Mechanism-based research on wind turbine blade icing often necessitates natural icing experiments in wind farms or the use of corresponding tools, such as icing wind tunnels. Neil Bose, by monitoring natural icing conditions in wind farms, concluded that glaze ice formed in freezing rain and drizzle has the greatest impact on wind turbines. He also found that icing is more severe on cloudy and rainy days, blade icing is more severe on the windward side than the leeward side, and ice accretion at the blade tip is significantly greater than at the blade root due to rotation [79]. Andrea Kraj et al., by controlling external environmental factors in an icing wind tunnel, obtained the temporal variation of blade icing and measured pressure changes on the wind turbine blade surface, providing new insights for de-icing [59]. Shi Lei et al. conducted icing research on two different airfoils and a wind turbine cylinder with a specific diameter in a low-speed icing wind tunnel. The icing tests of an aluminum cylinder, an NACA0018 airfoil, and an S809 airfoil were carried out, and the two-dimensional ice shape obtained by the test was compared with the two-dimensional ice shape obtained by the numerical simulation software [80]. Shu Lichun et al., through natural icing observations of a 2 KW wind turbine located in Xuefeng Mountain, Hunan, established a power calculation model for critical de-icing of wind turbine blades [81]. Hu Qin et al. combined artificial icing experiments and numerical simulations to study the icing of NE-100 horizontal axis wind turbine blades. This research obtained the degree of influence of different environmental factors on blade icing and depicted the morphology of blade ice accretion [73].

In the detection of icing on wind turbine blades, mechanism-based modeling has been widely applied and researched. However, currently, mechanism-based modeling often relies on high-precision sensors or icing wind tunnels. The common drawbacks of both are high cost and significant maintenance expenses. Moreover, blade icing is a highly complex process involving multiple factors such as temperature, humidity, wind speed, and turbine model. The interaction and coupling among these factors result in lower accuracy of models built on icing mechanisms. More seriously, with the passage of time and changes in environmental factors, sensors gradually age, making it difficult for mechanism-based icing detection models to accurately detect wind turbine blade icing. In the process of constructing mechanism-based models, it is often necessary to make assumptions about certain variables. Although such assumptions can simplify the model development process, they inevitably sacrifice model accuracy and limit their practical application value.

### 1.2.4 Data-Driven Methods

With the development of Industrial Internet of Things (IIoT) technology, an increasing amount of sensor data is being collected in industrial settings. Fault prediction based on collected sensor data has garnered significant attention from both academia and industry. Compared to mechanism-based modeling methods, data-driven modeling approaches treat the research object as a black box, de-emphasizing the need for and reliance on complex physical mechanisms. Data-driven methods directly start from blade icing data, establishing relationships between inputs and outputs. Given sufficient data, data-driven modeling methods can achieve excellent results.

Broadly, data-driven methods can be categorized into machine learning-based methods and deep learning-based methods. Machine learning-based methods first utilize human expert knowledge to construct various features from collected sensor data that can reflect the characteristics of the research object. Then, based on these constructed features, they build models with relatively simple structures, such as decision trees and multi-layer neural networks. Deep learning-based methods, on the other hand, directly model the collected sensor data, leveraging the model's own learning capabilities to learn potential patterns and features within the data. Machine learning-based methods often require considerable human effort and resources due to the need to rely on human expert knowledge to construct features. In contrast, deep learning-based methods, by eliminating the need for manual feature engineering, can improve modeling efficiency and reduce costs. It is worth noting that the term "feature" in the above context refers to indicators extracted from data that can be used to determine the state of an object. "Features" need to have a strong correlation with the research object. For instance, in facial recognition, features such as the position, shape, and relative distances of major facial organs can be constructed, and these features can then be used to build models to achieve facial recognition. In wind turbine blade icing detection, commonly extracted features primarily include statistical features in the time and frequency domains, energy spectra, and specific signals.

Data-driven methods for predicting wind turbine blade icing mainly involve six key steps:

1. Wind turbine blade icing scenario analysis. The essence of data-driven methods dictates their close binding to the research object. Therefore, it is essential to first fully understand the wind turbine icing background, clarify service objectives, and determine the functions and boundaries of the data-driven model to be built.
2. Wind turbine blade icing problem definition. After defining the business objectives for wind turbine blade icing, it is necessary to define the blade icing problem. In this text, we categorize the wind turbine blade icing problem as a classification problem, specifically, determining whether wind turbine blades are iced or not (two states).
3. Data preparation. Raw data collected from industrial sites often cannot be directly used for modeling due to issues with data quality and completeness. Therefore, before building a data-driven model, it is necessary to perform essential data quality checks to ensure that the data used for modeling aligns with the business objectives.
4. Model construction. Construct a data-driven model using the processed data. In industrial settings, the requirements for data-driven models are reliability, generalization ability, and interpretability.

5. Model evaluation. Evaluate the constructed model using appropriate evaluation metrics to ensure that the model can accurately generate information to support user decision-making.
6. Model deployment. Models validated offline need to be continuously iterated and updated in practical applications to improve productivity.

Skrimpas et al. identified wind turbine blade icing by combining nacelle vibration data and power curves. This method, belonging to machine learning-based approaches, analyzed wind speed-power datasets, extracted features such as power difference, nacelle lateral vibration, and operating month to build a KNN model. The effectiveness of this method was validated on icing data from the winters of 2012–2013 and 2013–2014 at a wind farm with 13 wind turbines [82]. Alfredo Arcos Jiménez et al. proposed a wind turbine blade icing detection model based on ultrasound and machine learning. This method presented effective methods for extracting and selecting ultrasonic signals and employed traditional machine learning methods such as decision trees, support vector machines (SVM), and KNN to build classification models [83]. Pandit et al., based on the characteristics of data collected by existing wind farm SCADA systems, proposed a set of data processing and feature mining methods suitable for wind turbine blade icing data. They used recurrent neural networks LSTM and GRU to build deep learning models [84]. Song Huan et al. proposed a wind turbine blade icing detection method based on image recognition. This method combined the Canny algorithm and salient object detection algorithms to detect blade icing by calculating pixel gradient changes around the blade [85]. Cai Fan et al. developed a fusion model based on historical data collected from wind farm SCADA systems, employing statistical learning methods and PCA, fLsm [86]. Zhang Dahai et al. proposed a wind turbine blade icing detection model combining random forest and SVM. This model used recursive feature elimination with random forest for feature selection from the original dataset, then trained an SVM with these selected features, and finally fused the SVM model and the random forest model [87]. Alvela et al., addressing the data variability across different types of wind turbines and operating conditions, proposed a wind turbine blade icing detection model based on transfer learning [88]. Jiang Guoqian et al. addressed the limitations of traditional statistical learning models and time-series models by proposing a spatio-temporal attention model to extract and assign weights to temporal–spatial features based on their contributions to blade icing detection. The effectiveness of this method was validated using data collected from actual wind farms [89].

In summary of the literature review and research status, it is evident that data-driven methods have achieved certain research outcomes and demonstrate superiority and great potential compared to mechanism-based modeling methods. However, data-driven wind turbine blade icing detection still faces the following challenges and limitations:

1. First, the data collected by wind farm SCADA systems is essentially time-series data. However, most current work has failed to fully investigate the temporal nature of operational data recorded during wind turbine operation. The loss of temporal information has resulted in the inability of the aforementioned studies to establish accurate icing prediction models.
2. Second, most studies have not considered data imbalance and the importance of certain sensor data. Because wind turbines operate in normal conditions most of the time, and blade icing occurs only for a small fraction of the time, SCADA recorded data exhibits significant imbalance. Simultaneously, SCADA systems collect a considerable amount of data, and how to effectively assess the importance of each sensor when building data-driven icing detection models has not been fully researched. Intuitively, temperature sensors may be more important in distinguishing between icing and non-icing conditions, but few studies provide a roadmap to evaluate whether this is the case.
3. Third, data-driven wind turbine blade icing detection is overly reliant on well-labeled data. However, labeling sensor data requires significant human and material resources. Few studies among the aforementioned research have addressed utilizing small amounts of labeled data to build wind turbine blade icing detection models.
4. Fourth, the above studies have all focused on centralized analysis and research of data collected by wind farm SCADA systems, without considering that in practice, owners of individual wind farms may be unwilling to share their collected sensor data with others due to data privacy concerns. Therefore, building data-driven learning models while effectively protecting the privacy of data owners is a direction that remains underemphasized.

## 1.3 Outline of This Book

This book, “Computational Methods for Blade Icing Detection of Wind Turbines”, presents a comprehensive exploration of advanced computational methodologies designed to address the critical challenge of blade icing detection on wind turbines. Recognizing the increasing reliance on wind energy and the detrimental effects of blade icing, this work delves into cutting-edge techniques to facilitate accurate and reliable detection, bridging the gap between theoretical innovation and practical application. Specifically, this book addresses the existing challenges and limitations in the detection of wind turbine blade icing through the introduction of novel computational methodologies that integrate aspects of temporal convolutional neural network, graph convolutional neural networks, and federated convolutional neural networks. The book is structured into three distinct parts, each focusing on a specific methodological approach, and encompasses eleven chapters that span the breadth of computational techniques from deep learning to graph neural networks to federated learning, all within the context of this significant challenge. Ultimately, this work aims to serve as an invaluable resource for the academic community as well as for the engineers and practitioners working to advance wind energy and to improve its sustainability in cold and remote climates.

Part I lays the foundation by delving into Temporal Convolutional Neural Network (TCN) Based Methods. This section focuses on the potential of advanced TCN architectures for providing robust, precise, and practical solutions for blade icing detection. Each chapter in this part introduces a novel TCN-based approach: Chapter 2 details the design and evaluation of the Attention-Enhanced Temporal Convolutional Neural Network (TACNN) which uses a temporal attention mechanism to better discern the importance of individual sensors and time steps in time-series data. Chapter 3 elaborates on a Multilevel Convolutional Recurrent Neural Network (MCRNN) which leverages wavelet decomposition for multi-scale feature extraction and novel recurrent architecture to capture spatio-temporal dependencies. Finally, Chapter 4 introduces a novel Deep Class-Imbalanced Semi-Supervised (DCISS) model, which addresses data imbalances common in real-world datasets and allows for learning from both labeled and unlabeled data. These chapters collectively explore the power of deep learning for automatic feature extraction, emphasizing the importance of temporal dynamics in achieving robust and reliable performance.

Part II transitions to Graph Neural Network-Based Methods, acknowledging the inherent interconnectedness of wind turbine sensor networks. This section emphasizes the value of graph neural networks (GNNs) for modeling intricate sensor relationships and extracting meaningful information for blade icing detection. Chapter 5 presents an Attention-Enhanced Graph Convolutional Neural Network (GTAN) that explicitly models spatial-temporal dependencies for improved detection. Chapter 6 introduces a Wavelet-Empowered Graph Convolutional Neural Network (MWGCN) that uses wavelet analysis for multiscale feature extraction combined with temporal graph neural networks, further enhancing the capacity to harness temporal data. Finally, Chapter 7 explores a Class-Imbalanced Graph Temporal Convolutional Neural Network which utilizes a distance-based classifier to mitigate data imbalances. This section highlights the potential for GNNs to enhance model complexity and extract meaningful features through various attention and graph-based modeling approaches.

Part III pivots to Federated Convolutional Neural Network-Based Methods, focusing on the transformative potential of federated learning (FL). This section examines methods for collaborative model training without necessitating direct sharing of raw data, recognizing the increasing need for privacy-preserving and distributed data processing techniques. Chapter 8 explores a Split Heterogeneous Federated Convolutional Neural Network (BiFL) for blade icing detection, specifically designed to address data and model heterogeneity among wind turbines, utilizing split learning and knowledge distillation. Chapter 9 presents a Data-Efficient Heterogeneous Federated Convolutional Neural Network (FedBIP), which leverages feature selection and knowledge distillation within the federated learning paradigm to enhance performance, training, and prediction while mitigating data scarcity. Chapter 10 introduces a Blockchain-Empowered Clustered Federated Learning Model for blade icing estimation on IoT-enabled Wind Turbines, which explores blockchain technology within a clustered federated learning framework, ensuring secure and reliable model training. This section underscores how federated learning can facilitate collaborative model training, protect data privacy, and improve model performance through knowledge transfer, novel security measures, and effective techniques for class imbalance.

The final chapter, Chapter 11, serves to provide concluding remarks, summarizing the key findings of each part, discussing their practical implications, and outlining prospective avenues for future investigation. This chapter aims to both demonstrate the book's overall contribution to the field of blade icing detection and to chart a course for future explorations into this important area.

## References


1. USEIA (2019). https://www.eia.gov/outlooks/ieo/pdf/ieo2019.pdf
2. USEIA (2021). https://www.eia.gov/totalenergy/data/monthly/
3. Solar (2021). https://solar.in-en.com/html/solar-2363068.shtml
4. BJX (2021). http://news.bjx.com.cn/
5. Nikolaus E (2014) The right size matters: investigating the offshore wind turbine market equilibrium. Energy 68:910–921
6. UpWind (2011) Design limits and solutions for very large wind turbines–a 20 mw turbine is feasible
7. Erich H (2013) Wind turbines: fundamentals, technologies, application, economics. Springer Science and Business Media, USA
8. Tummala A, Velamati RK, Sinha DK, Indraja V, Hari Krishna V (2016) A review on small scale wind turbines. Renew. Sustain. Energy Rev. **56**, 1351–1371
9. Lin W, Xiongwei L, Athanasios K (2016) State of the art in the aeroelasticity of wind turbine blades: aeroelastic modelling. Renew. Sustain. Energy Rev. 64:195–210
10. Jamieson P, Branney M (2014) Structural considerations of a 20mw multi-rotor wind energy system. J Phys: Conf Ser 555:012013. Oldenburg, Germany. IOP Publishing, IOP Publishing
11. Olivier P, Adrian I (2011) Anti-icing and de-icing techniques for wind turbines: critical review. Cold Regions Sci Technol 65(1):88–96
12. Ronsten G, Wallenius T, Hulkkonen M, Baring-Gould I, Cattin R, Durstewitz M, Krenn A, Laakso T, Lacroix A, Tallhaug L et al (2012) IEA wind task 19: State-of-the-art of wind energy in cold climates. Technical report, IEA wind, USA, Technical report
13. IEA Wind. IEA. IEA wind task 19: wind energy projects in cold climates. Technical report, IEA Wind, USA (2016)
14. Jordaens PJ, Krenn A, Wadham-Gagnon M, Davis N, Cattin R (2016) IEA wind task 19—available technologies report of wind energy in cold climates
15. Lichun S, Hantao L, Qin H, Xingliang J, Gang Q, Ghyslaine MC, Hang Y (2018) Study of ice accretion feature and power characteristics of wind turbines at natural icing environment. Cold Regions Sci Technol 147:45–54
16. Dierer S, Oechslin R, Cattin R (2011) Wind turbines in icing conditions: performance and prediction. Adv Sci Res 6:245
17. René C (2012) Icing of wind turbines. Vindforsk projects, a survey of the development and research needs. Elforsk Rep. 12:13
18. Shilin S, Tianyang W, Fulei C (2022) In-situ condition monitoring of wind turbine blades: a critical and systematic review of techniques, challenges, and futures. Renew Sustain Energy Rev 160:112326

19. Anne B, Garry T, Helge G, Per V (2005) Ice loading on Danish wind turbines: part 1: dynamic model tests. Cold Regions Sci Technol 41(1):1–23
20. Pinar-Pérez JM, Márquez FPG (2018) Managing costs and review for icing problems. Springer International Publishing, Cham, pp 97–109
21. Pérez JMP, Márquez FPG, Hernández DR (2016) Economic viability analysis for icing blades detection in wind turbines. J Clean Product 135:1150–1160
22. Ville L, Simo R, Matthew W-G, Klaus S, Wolfgang M, Danela J (2016) Fatigue loads of iced turbines: two case studies. J Wind Eng Ind Aerodyn 158:37–50
23. Barber S, Wang Y, Jafari S, Chokani N, Abhari RS (2011) The impact of ice formation on wind turbine performance and aerodynamics. J Solar Energy Eng 133(1)
24. Hellström E (2013) Development of a model for estimation of wind farm production losses due to icing. Energisystem
25. Ryan Kilpatrick ME, Eng P (2017) Effect of cold climate on wind energy production in Canada (2010–2016)
26. Peter B, Jose P, Sven S (2017) Effect of icing roughness on wind turbine power production. Wind Energy 20(4):601–617
27. Etemaddar M, Laver Hansen MO, Moan T (2014) Wind turbine aerodynamic response under atmospheric icing conditions. Wind Energy 17(2):241–265
28. Homola MC, Virk MS, Nicklasson PJ, Sundsbø PA (2012) Performance losses due to ice accretion for a 5 mw wind turbine. Wind Energy 15(3):379–389
29. Wang X, Bibeau EL, Naterer GF (2007) Experimental investigation of energy losses due to icing of a wind turbine. In: Cen K, Chi Y, Wang F (eds), Challenges of power engineering and environment, Berlin, Heidelberg. Springer Berlin Heidelberg, pp 1143–1147
30. Jing Y, Wei Yu, Julien C, Alain F, Slavica A (2015) Coupled atmospheric-ice load model for evaluation of wind plant power loss. J Appll Meteorol Climatol 54(6):1142–1161
31. Fayçal L, Guy F, Robert B, Jean P, Christian M (2014) Atmospheric icing impact on wind turbine production. Cold Regions Sci Technol 100:36–49
32. Wadham-Gagnon M, Bolduc D, Boucher B, Camion A, Petersen J, Friedrich H (2013) Ice profile classification based on iso 12494. In: Proceedings of the WinterWind, p 48
33. Eleni DC, Athanasios TI, Dionissios MP (2012) Evaluation of the turbulence models for the simulation of the flow over a national advisory committee for aeronautics (NACA) 0012 Airfoil. J Mech Eng Res 4(3):100–111
34. Durga Charan Panigrahi and Devi Prasad Mishra (2014) CFD simulations for the selection of an appropriate blade profile for improving energy efficiency in axial flow mine ventilation fans. J Sustain Mining 13(1):15–21
35. Dongli M, Yanping Z, Yuhang Q, Guanxiong L (2015) Effects of relative thickness on aerodynamic characteristics of airfoil at a low reynolds number. Chinese J Aeronaut 28(4):1003–1015
36. Fernando V, Marcelo R, Adrian I (2012) Numerical study of flow around iced wind turbine airfoil. Eng Appl Comput Fluid Mech 6(1):39–45
37. Timmer WA, van Rooij RPJOM (2003) Summary of the delft university wind turbine dedicated airfoils. J Sol Energy Eng 125(4):488–496
38. Van Rooij R, Timmer WA (2003) Roughness sensitivity considerations for thick rotor blade airfoils. J Sol Energy Eng 125(4):468–478
39. Llorente E, Gorostidi A, Jacobs M, Timmer WA, Munduate X, Pires O (2014) Wind tunnel tests of wind turbine airfoils at high reynolds numbers. J Phys: Conf Ser 524:012012
40. Ibrahim GM, Pope K, Muzychka YS (2018) Effects of blade design on ice accretion for horizontal axis wind turbines. J Wind Eng Ind Aerodyn 173:39–52
41. Jia Yi Jin and Muhammad Shakeel Virk (2019) Study of ice accretion and icing effects on aerodynamic characteristics of du96 wind turbine blade profile. Cold Regions Sci Technol 160:119–127
42. Kunberger V, Weigand B (2024) Binary collisions of supercooled water droplets. Atomization Sprays 34(12)
43. Liangquan H, Xiaocheng Z, Chenxing H, Jinge C, Zhaohui D (2017) Calculation of the water droplets local collection efficiency on the wind turbines blade. J Energy Res Technol 139(5):051211

44. Haifeng Q, Shinan C, Yinglin Y (2023) Numerical study of mixed phase ice accumulation in aero-engine inlet system. Appl Thermal Eng 231:120909
45. Baumert A, Bansmer S, Trontin P, Villedieu P (2018) Experimental and numerical investigations on aircraft icing at mixed phase conditions. Int J Heat and Mass Trans 123:957–978
46. Zhao B, Guo P (2022) Wind turbine blade icing detection with scada data. In: 2022 34th Chinese control and decision conference (CCDC). IEEE, pp 5256–5261
47. Han Y (2011) Theoretical and experimental study of scaling methods for rotor blade ice accretion testing. PhD thesis, Pennsylvania State University, USA
48. Jiang W (2020) Effect of blade surface roughness on wind energy utilization coefficient. In: AIP conference proceedings, vol 2238. AIP Publishing
49. Francesco C, Alberto G (2023) Numerical simulations of ice accretion on wind turbine blades: are performance losses due to ice shape or surface roughness? Wind Energy Sci 8(3):341–362
50. Rustem M, Zhandos B, Saltanat B, Aleksandar G (2021) Numerical simulations on static vertical axis wind turbine blade icing. Renew Energy 170:997–1007
51. Gao L (2019) Experimental investigations on wind turbine icing physics and anti-/de-icing technology
52. Albers A (2011) Summary of a technical validation of enercon's rotor blade de-icing system. Deutsche WindGuard Consulting GmbH, Varel, Germany, Report No PP11035
53. Oloufemi F, Zoé F, Hussein I, Adrian I, Jean P, Christian M (2016) Ice protection systems for wind turbines in cold climate: characteristics, comparisons and analysis. Renew Sustain Energy Rev 65:662–675
54. Mayer C, Ilinca A, Fortin G, Perron J et al (2007) Wind tunnel study of the electro-thermal de-icing of wind turbine blades. In: The seventeenth international offshore and polar engineering conference, Lisbon, Portugal. International Society of Offshore and Polar Engineers, ISOPE
55. Jinsheng C, Yongqiang T, Xuanshi M, Xuzhao H, Duo Z, Haiyang H (2017) An experimental study of icing control using DBD plasma actuator. Exp Fluids 58(8):1–8
56. Oloufemi F, Hussein I, Adrian I, Jean P (2016) Experimental investigation of power requirements for wind turbines electrothermal anti-icing systems. Wind Turbines-Design, Control and Applications
57. Gao L, Liu Y, Kolbakir C, Hu H (2018) An experimental investigation on an electric-thermal strategy for wind turbines icing mitigation. In: 2018 atmospheric and space environments conference, Atlanta, Georgia. AIAA, p 3658
58. Mohseni M, Amirfazli A (2013) A novel electro-thermal anti-icing system for fiber-reinforced polymer composite airfoils. Cold Regions Sci Technol 87:47–58
59. Kraj AG, Bibeau EL (2010) Phases of icing on wind turbine blades characterized by ice accumulation. Renew Energy 35(5):966–972
60. Wang Z-J, Kwon D-J, De Vries KL, Park J-M (2015) Frost formation and anti-icing performance of a hydrophobic coating on aluminum. Exp Thermal Fluid Sci 60:132–137
61. Chaoyi P, Suli X, Zhiqing Y, Jiayu X, Chunqi W, Jingcheng Z (2012) Preparation and anti-icing of superhydrophobic pvdf coating on a wind turbine blade. Appl Surf Sci 259:764–768
62. Qi L, Ying Y, Meng H, Yuanxiang Z, Yingyan L, Xidong L (2015) Durability of a lubricant-infused electrospray silicon rubber surface as an anti-icing coating. Appl Surf Sci 346:68–76
63. Adrián L-V, André MD (2016) Flame-sprayed coatings as de-icing elements for fiber-reinforced polymer composite structures: modeling and experimentation. Int J Heat Mass Trans 97:56–65
64. Liangquan H, Xiaocheng Z, Chenxing H, Jinge C, Zhaohui D (2017) Wind turbines ice distribution and load response under icing conditions. Renew Energy 113:608–619
65. Congbo Y, Zhendong Z, Zhenjun W, Hui G (2016) Numerical simulation and experimental validation of ultrasonic de-icing system for wind turbine blade. Appl Acoust 114:19–26
66. Hossein H, Graham E, Caumaghen S, Vassilios K, Yoann L, Jasmin S, Cem S, Tat-Hean G (2016) Modelling and empirical development of an anti/de-icing approach for wind turbine blades through superposition of different types of vibration. Cold Regions Sci Technol 128:1–12
67. Seifert H (2005) Technical requirements for rotor blades operating in cold climate. DEWI-Magazin

68. Hossein H, Liang C, Haitao Z, Vassilios K, Cem S, Tat-Hean G (2015) A dual de-icing system for wind turbine blades combining high-power ultrasonic guided waves and low-frequency forced vibrations. Renew Energy 83:859–870
69. Makkonen L, Laakso T, Marjaniemi M, Finstad KJ (2001) Modelling and prevention of ice accretion on wind turbines. Wind Eng 25(1):3–21
70. Saleh SA, Ahshan R, Moloney CR (2012) Wavelet-based signal processing method for detecting ice accretion on wind turbines. IEEE Trans Sustain Energy 3(3):585–597
71. Davis NN, Byrkjedal Ø, Hahmann AN, Clausen N-E, Žagar M (2016) Ice detection on wind turbines using the observed power curve. Wind Energy 19(6):999–1010
72. Zhengzhi W, Chunling Z (2018) Numerical simulation for in-cloud icing of three-dimensional wind turbine blades. Simulation 94(1):31–41
73. Qin H, Hang Y, Xingliang J, Lichun S, Xin Y (2022) Investigation on one-step preparation and anti-icing experiments of robust super-hydrophobic surface on wind turbine blades. Cold Regions Sci Technol 195:103484
74. Ivan K, Sergey D, Maxim G, Sergey K, Vadim L, Grigoriy G, Vladislav Z, Valery O (2021) Optical methods for measuring icing of wind turbine blades. Energies 14(20):6485
75. Ping F, Masoud F (2010) A CFD approach for modeling the rime-ice accretion process on a horizontal-axis wind turbine. J Wind Eng Ind Aerodyn 98(4–5):181–188
76. Homola MC, Virk MS, Wallenius T, Nicklasson PJ, Sundsbø PA (2010) Effect of atmospheric temperature and droplet size variation on ice accretion of wind turbine blades. J Wind Eng Ind Aerodyn 98(12):724–729
77. Homola MC, Wallenius T, Makkonen L, Nicklasson PJ, Sundsbø PA (2010) The relationship between chord length and rime icing on wind turbines. Wind Energy 13(7):627–632
78. Sierikova O, Strelnikova E, Degtyariov K (2022) Numerical simulation of strength and aerodynamic characteristics of small wind turbine blades. In: Conference on integrated computer technologies in mechanical engineering–synergetic engineering. Springer, pp 357–370
79. Neil B (1992) Icing on a small horizontal-axis wind turbine-part 1: glaze ice profiles. J Wind Eng Ind Aerodyn 45(1):75–85
80. Lei S, Fang F, Wenfeng G, Yan L (2021) Research and development of a small-scale icing wind tunnel test system for blade airfoil icing characteristics. Int J Rotat Mach 1:5598859
81. Lichun S, Gang Q, Qin H, Xingliang J, Ghyslaine MC, Hang Y (2018) Numerical and field experimental investigation of wind turbine dynamic de-icing process. J Wind Eng Ind Aerodyn 175:90–99
82. Skrimpas GA, Kleani K, Mijatovic N, Sweeney CW, Jensen BB, Holboell J (2016) Detection of icing on wind turbine blades by means of vibration and power curve analysis. Wind Energy 19(10):1819–1832
83. Jiménez AA, Márquez FPG, Moraleda VB, Muñoz CQG (2019) Linear and nonlinear features and machine learning for wind turbine blade ice detection and diagnosis. Renew Energy 132:1034–1048
84. Ravi P, Davide A, Jiarong H, David I, Matilde S (2023) Scada data for wind turbine data-driven condition/performance monitoring: a review on state-of-art, challenges and future trends. Wind Eng 47(2):422–441
85. Huan S, Yongguang H (2023) Detecting emerging frost crystals on leaf surfaces using microscopic imaging. Comput Electron Agric 210:107931
86. Fan C, Yuesong J, Wanqing S, Kai-Hung L, Tongbo Z (2024) Short-term wind turbine blade icing wind power prediction based on PCA-FLSM. Energies 17(6):1335
87. Dahai Z, Liyang Q, Baijin M, Can H, Bin H, Yulin S (2018) A data-driven design for fault detection of wind turbines using random forests and xgboost. IEEE Access 6:21020–21031
88. Nieto MTA, Gelbhardt H, Ohlendorf J-H, Thoben K-D (2022) Detecting ice on wind turbine rotor blades: towards deep transfer learning for image data. In: International conference on system-integrated intelligence. Springer, pp 574–582
89. Guoqian J, Ruxu Y, Qun H, Ping X, Xiaoli L (2023) Imbalanced learning for wind turbine blade icing detection via spatio-temporal attention model with a self-adaptive weight loss function. Exp Syst Appl 229:120428

# Part I
# Temporal Convolutional Neural Network-Based Methods

As highlighted in the preceding chapter, the reliable detection of blade icing on wind turbines remains a critical hurdle to the consistent and cost-effective generation of wind energy. While traditional methodologies and physics-based models offer valuable insights, their practical application is often hampered by substantial costs, scalability limitations, or a heavy dependence on expert domain knowledge. Data-driven approaches, and especially those leveraging the power of Temporal Convolutional Neural Networks (TCNs), have emerged as compelling alternatives for confronting this intricate challenge. TCNs, with their inherent capability to automatically extract complex temporal features from unprocessed, high-dimensional sensor data, and their aptitude for modeling intricate, non-linear time-series relationships, are exceptionally well-suited for the task of wind turbine blade icing detection. Their architecture, optimized for processing sequential data, allows for efficient parallel computation and long-range dependency capture, making them particularly advantageous for analyzing the time-series data from wind turbine SCADA systems. By circumventing the need for manual feature engineering, TCNs empower the network to autonomously learn crucial temporal characteristics directly from data. This enables the development of highly accurate and efficient models, ultimately bolstering the operational safety and diminishing the maintenance expenditures associated with wind farms.

Part I is dedicated to exploring a suite of innovative Temporal Convolutional Neural Network (TCN)-based methodologies meticulously designed to address the unique challenges inherent in wind turbine blade icing detection. These methods rigorously investigate the potential of advanced TCN architectures to deliver more robust, precise, and practically viable solutions. Leveraging the inherent strengths of convolutional layers within the TCN framework, alongside recurrent components and sophisticated attention mechanisms, these architectures exhibit a remarkable proficiency in processing the complex time-series data streams generated by wind farm SCADA systems. For instance, Chap. 2 introduces the Attention-Enhanced Temporal Convolutional Neural Network (TACNN), a novel TCN-based model designed to address the challenge of discerning the importance of different sensors and time steps. TACNN innovatively incorporates a temporal attention mechanism before the TCN convolutional layers, dynamically weighting sensor and time-step significance to enhance feature extraction and capture crucial temporal dependencies within the time-series data. Building upon this foundation, Chap. 3 proposes a Wavelet-Enhanced Temporal Convolutional Neural Network, another innovative TCN architecture, engineered to capture comprehensive features across both time and frequency domains. This method utilizes Multilevel Discrete Wavelet Decomposition (MDWD) to extract multiscale features, processed by a Multilevel Convolutional Recurrent Neural Network (MCRNN) integrated within the TCN framework, to capture spatial and temporal dependencies, effectively handling the complexity of wind turbine signals. Further addressing practical deployment limitations, Chap. 4 presents the Class-Imbalanced Semi-supervised Temporal Neural Network (DCISS), a specialized TCN-derived model, specifically designed to tackle labeled data scarcity and class imbalance. DCISS synergistically combines semi-supervised and class-imbalanced learning using a prototypical network architecture, built upon TCN principles, enabling learning from both labeled and unlabeled data, thereby enhancing data efficiency in real-world scenarios. Each method is meticulously crafted to tackle a distinct challenge within the field and is purposefully designed to contribute significantly to the advancement of icing detection capabilities using the power of TCNs.

The central objective of this part is to furnish a comprehensive understanding of how Temporal Convolutional Neural Network methodologies can be effectively harnessed to enhance the precision of blade icing estimation. These TCN-based methods facilitate end-to-end learning directly from raw sensor data and are engineered to discern discriminative temporal features and capture intricate interdependencies within the time-series data. By employing cutting-edge techniques, including temporal attention mechanisms and wavelet transformations, as exemplified by TACNN and the Wavelet-Enhanced TCNN, these TCN models are adept at processing data characterized by substantial noise levels, inherent non-linearity, and temporal complexity—conditions frequently encountered in the real-world operational environments of wind turbines. Moreover, these methodologies introduce novel strategies to mitigate practical limitations, such as pronounced class imbalance and the scarcity of labeled data, as demonstrated by the DCISS approach, all within the context of TCN architectures.

Each chapter encompassed within this part meticulously details a unique TCN-based deep learning approach, substantiated by rigorous experimental evaluations conducted on publicly accessible benchmark datasets, as well as real-world Supervisory Control and Data Acquisition (SCADA) data harvested from operational wind turbine systems. The experimental findings comprehensively presented throughout this part emphatically underscore the considerable potential and practical applicability of Temporal Convolutional Neural Networks within the context of blade icing detection. This lays a robust foundation for the realization of more dependable and efficient wind energy operations, especially in regions prone to cold climates. This research significantly enriches the broader understanding of data-driven modeling strategies, specifically those leveraging TCNs, tailored for deployment in complex and noisy operational settings.

# Chapter 2
# Attention-Enhanced Temporal Convolutional Neural Network for Blade Icing Detection

## 2.1 Introduction

Wind power, celebrated for its abundance and sustainability, has become a cornerstone of the global transition towards renewable energy sources [1]. To maintain a consistent and reliable energy supply, wind turbines must function effectively across diverse and fluctuating climatic conditions. Among these, blade icing, which is caused by the accumulation of ice during winter, significantly alters their aerodynamic properties and torque characteristics, leading to diminished power output and increased fatigue loads [2]. Therefore, a timely and accurate detection of blade icing has become a critical necessity for ensuring the efficient and safe operation of modern wind farms, and as such, the development of effective computational methods for blade icing detection has become a crucial research direction [3, 4].

Traditionally, efforts to mitigate blade icing have fallen into three primary categories: passive, active, and hybrid approaches. *Passive methods*, like specialized coatings [4, 5], attempt to prevent ice formation but are limited by their inability to eliminate it entirely. They may also increase operational costs due to material changes on blade surfaces. *Active methods* use external systems to eliminate ice but often cause mechanical failures and require extra energy [6]. Hybrid methods integrate aspects of both, and still often have limitations in terms of costs and internal inconsistencies. Moreover, all of these methods also may necessitate mechanical replacements.

In response to these limitations, research has shifted towards methods using turbine operational measurements such as model-based and data-driven. Model-based approaches use mathematical and numerical models with the aid of domain knowledge [7], yet they are reliant on external tools and domain knowledge, limiting their practical applicability. Data-driven methods, using machine and deep learning, offer a more flexible alternative by leveraging readily available sensor data. They can also be more cost-effective since they reduce the need for external tools and specialized



X. Cheng et al., *Computational Methods for Blade Icing Detection of Wind Turbines*, Engineering Applications of Computational Methods 24,
https://doi.org/10.1007/978-981-96-6763-5_2

expertise. Deep learning methods [6, 8, 9] have the additional advantage of automating the feature extraction, bypassing the need for hand-crafted features of shallow learning-based methods [10–14].

However, despite these advantages, deep learning still faces several critical challenges when applied to the problem of wind turbine blade icing detection. First, raw sensor data from wind turbines is inherently characterized by high levels of non-linearity and non-stationarity. These characteristics arise from the constantly shifting environmental conditions, including variations in temperature, wind speed, and humidity. These fluctuations lead to sensor data that cannot be accurately modeled by linear methods and that shifts over time, creating major obstacles for effective feature extraction by deep learning models. Furthermore, it makes it more difficult to apply methods designed for the analysis of general time-series data to the specific problem of blade icing detection. Second, the real-world data obtained for blade icing detection is highly imbalanced. Wind turbines operate under normal (non-icing) conditions the majority of the time, with icing events occurring only sporadically. This results in a significant class imbalance, where the number of non-icing samples drastically outnumbers icing samples. Such data imbalance can cause deep learning models to become biased towards the majority class (non-icing samples), diminishing their ability to effectively detect the features associated with the minority class (icing samples), and ultimately, compromising the detection accuracy. Third, the need for early and predictive detection of icing is paramount because proactive action taken at an early stage can help in reducing icing induced damage in the wind turbines. However, most current deep learning approaches are designed for making predictions after the icing event has already occurred. Achieving early prediction of blade icing requires models with a high degree of temporal awareness, capable of identifying subtle patterns associated with the onset of icing and anticipating icing trends before they fully materialize. Finally, and perhaps most importantly, existing time-series classification methods often lack an explicit mechanism to evaluate the importance of individual sensors and timesteps. Consequently, such methods tend to treat all sensors as equally important and neglect the temporal context in the data. Given that some sensors, such as temperature sensors, may be more directly correlated with blade icing than others, and that specific time steps might hold more relevant information, the inability to differentiate sensor and timestep significance limits model performance and their ability to focus on the most relevant information.

The task of detecting icing conditions is often framed as a Time Series Classification (TSC) problem. Existing TSC methods can be broadly categorized into distance-based, feature-based, and deep learning-based methods. Distance-based methods [15] use discrete SVM and warping distances, which are computationally intensive with limited adaptability for time series. Feature-based methods utilize patterns extracted from time-series data [16–18], which suffer from a reliance on manual feature engineering. While deep learning has alleviated some limitations, its use in detecting blade icing is still limited. Some examples of these deep learning approaches are baseline models by Wang et al. [19], a hybrid LSTM-FCN architecture introduced by Fazle et al. [20], a spectral branch with LSTM-FCN by Cheng et al. [21] and model by Cheng et al. combining dense connections and

CNNs [22]. Further, TapNet proposed by Zhang et al. demonstrated competitive performance [23]. However, these methods lack the ability to identify the key sensors and timesteps in raw time-series data.

To overcome these limitations, we introduce the Temporal Attention-based Convolutional Neural Network (TACNN), which leverages CNNs for feature extraction and a Temporal Attention (TA) module that allows the model to focus on the most pertinent information based on sensor and time step significance. Using an LSTM network before the convolutional layer, allows the model to capture long-term dependencies, enabling it for early-stage predictions. Thus, our proposed TACNN model overcomes the limitations of traditional CNN-based approaches and provides a more accurate and robust detection method.

This research offers several notable contributions:

(1) The study presents TACNN, a novel deep learning network that integrates a TA module into a conventional CNN. TACNN enhances feature extraction by discerning the importance of sensors and time steps, addressing the limitations of traditional CNNs that treat all sensors and timesteps equally. An end-to-end framework based on TACNN is developed for icing detection on wind turbine blades. This framework efficiently processes highly imbalanced sensor data while enabling automatic discriminative feature learning and accurate identification of icing conditions.
(2) The performance of TACNN is assessed using ten benchmark datasets of multivariate TSC and real supervisory control and data acquisition (SCADA) data from three wind turbines. Comparative analysis against state-of-the-art TSC methods across the benchmark datasets demonstrates the superior performance of TACNN. Additionally, comparisons with baseline networks and other attention modules in the SCADA data highlight the superiority and significance of the proposed model. The generalizability and practicality of TACNN are further confirmed through online estimation testing.

The rest of this research is structured as follows. Section II reviews the related work for wind turbine blade icing detection and TSC. The proposed TACNN is presented in Section III. The performance of the proposed approach is evaluated in Section IV. Section V emphasizes the conclusions and future work.

## 2.2 Methodology

### *2.2.1 Overview*

The proposed method comprises three key components: data preprocessing, a temporal attention convolutional neural network (TACNN), and icing detection, as illustrated in Fig. 2.1. Ensuring the quality of data is crucial for the performance of data-driven models. Therefore, researchers typically begin with data cleaning, preprocessing, and correlation analysis to reduce uncertainties related to data quality.

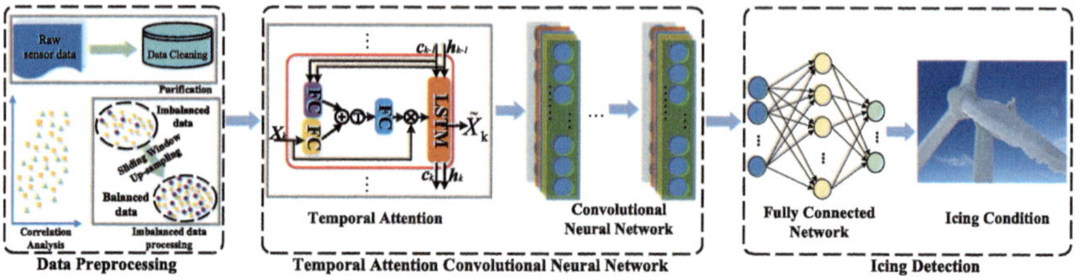


**Fig. 2.1** Framework of wind turbine blade icing detection. The photo of frozen wind turbine blades is from [24]

As wind turbines mainly operate in ice-free conditions, addressing imbalanced data becomes essential. In the proposed approach, processed sensor data are fed into the TACNN model. These data undergo preprocessing by the temporal attention (TA) module, which identifies key sensors and significant time steps. The weighted sensor data are then passed to the convolutional neural network (CNN) for feature extraction. Finally, the extracted features are used to compute the probability of icing via the fully connected (FC) network.

### 2.2.2 Data Preprocessing

As depicted in Fig. 2.1, data preprocessing includes several critical steps: data cleaning, correlation analysis, and managing imbalanced data. Cleaning the raw sensor data is vital to minimize the impact of outliers and noise, ensuring the quality and reliability of the subsequent analysis. Pearson correlation analysis is used to understand the relationships between different sensors, which helps in reducing redundant information and enhancing the efficiency of the data processing. Addressing class imbalance is a significant challenge in this research. Traditional re-balancing methods include undersampling the majority class and oversampling the minority class. However, undersampling can lead to the loss of valuable information, and improper oversampling can cause overfitting issues [28]. Another approach involves using specialized loss functions, such as the focal loss [29], which aim to improve model performance by focusing on hard-to-classify samples. However, this approach requires careful optimization of hyperparameters and has not always resulted in significant performance gains in practical applications. To effectively manage data imbalance, this study employs a sliding window upsampling (SWU) technique [3]. The SWU method is both simple and effective. For non-icing data, a non-overlapping sliding window is used to partition the data into segments of a predefined size, while an overlapping sliding window is applied to the icing data. As illustrated in Fig. 2.2, this approach collects more non-icing data points than icing data points (eight points for non-icing compared to five points for icing). By using SWU, four samples can be generated from non-icing data with a non-overlapping sliding window, and four

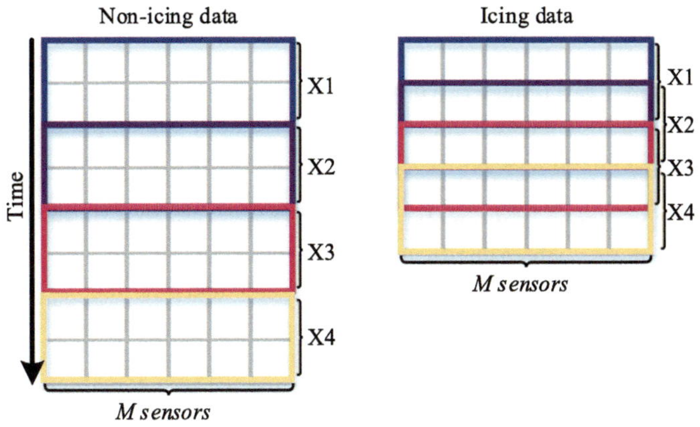


**Fig. 2.2** An illustration example of SWU

samples from icing data with an overlapping sliding window, assuming a window size of two. This technique ensures that both classes are adequately represented in the training data, improving the model's ability to accurately detect icing conditions.

### *2.2.3 Temporal Attention Convolutional Neural Network*

CNNs are employed for automatic feature learning due to the diverse environmental conditions under which wind turbines operate. Previous research has primarily focused on improving the feature learning capabilities of CNNs through various structural design enhancements. Despite these advancements, traditional CNNs often struggle to adaptively emphasize the most relevant distinguishing features while effectively ignoring those that are less relevant or irrelevant. This limitation can hinder the model's performance in accurately detecting complex patterns such as icing conditions on wind turbine blades. To address this challenge, attention mechanisms are typically utilized. These mechanisms enable the model to dynamically focus on the most informative parts of the input data, thereby enhancing the overall accuracy and robustness of the feature learning proccss.

Attention is a key aspect of human perception. The squeeze-and-excitation (SE) module stands out as the pioneering attention mechanism integrated into an FCN model and applied to Time Series Classification (TSC) [20]. Additionally, other attention modules like the convolutional block attention module (CBAM) [25] and global contexts (GC) [26] have been specifically developed for applications in computer vision. These attention mechanisms share a common characteristic: they re-weight the features extracted by a CNN to highlight their importance, thereby enhancing the model's focus on crucial aspects of the data. Typically, these modules are applied after the CNN layers to adjust the significance of the features. While these attention mechanisms can significantly improve the performance of CNNs by emphasizing critical features, they fall short in recognizing the importance of each individual sensor and timestep in raw time-series data. This limitation is particularly significant for tasks like wind turbine blade icing detection, where understanding the temporal and sensor-specific nuances is essential for accurate prediction and analysis.

To address attention-related challenges, our study introduces a unique Temporal Attention (TA) module, positioned distinctively before the CNN layer. Unlike conventional attention mechanisms such as SE, CBAM, or GC, our TA module offers direct computation of the significance of each sensor and timestep.

To ascertain the significance of individual sensors and time steps, we exploit the cell state $c_{t-1}$ and hidden state $h_{t-1}$ from previous timesteps within the LSTM framework, as outlined by Qin et al. [27]. Illustrated in Fig. 2.1's middle panel, these states ($c_{t-1}$ and $h_{t-1}$) undergo transformation through a fully connected (FC) layer. Likewise, the input undergoes a similar transformation, as shown in the figure. These encoded inputs and states then undergo diverse operations like addition, hyperbolic tangent, and FC transformations before entering the LSTM module.

Given $N$ sensors, $T$ timesteps, and $M$ hidden units in LSTM, the shape of the purple FC is denoted as $W_p \in \mathbb{R}^{2M \times T}$, the yellow FC as $W_y \in \mathbb{R}^{T \times T}$, and the blue FC as $W_b \in \mathbb{R}^{T \times 1}$. The subscripts $p$, $y$, and $b$ correspond to the colors of the three FC layers, as depicted in Fig. 2.1. The calculations of these FC layers can be expressed as follows:

$$
\begin{aligned}
z_p &= W_p([h_{t-1}; c_{t-1}]),\\
z_y &= W_y(X_k),\\
z_b &= W_b(tan(z_p + z_y)),
\end{aligned}
\tag{2.1}
$$

where $z_p$, $z_y$, and $z_b$ donate the output of the three FC layer. Then, the importance of each sensor can be computed as

$$
\alpha_k = softmax(z_b) = \frac{exp(z_b)}{\sum_{j=1}^{n} exp(z_b^j)},
\tag{2.2}
$$

where $\alpha_k$ is the weight for each timestep in $X_k$. Thus, the input for each timestep can be represented as

$$
\widetilde{X_k^t} = \alpha_k * X_k^t = (\alpha_k^1 X_k^1, \alpha_k^2 X_k^2, \cdots, \alpha_k^n X_k^n).
\tag{2.3}
$$

The entire time series undergoes weighting by LSTM before being passed into the CNN for feature extraction. Each CNN layer consists of a basic convolutional (CONV) layer, followed by batch normalization (BN), and rectified linear unit (RELU) operations stacked sequentially. Specifically, we employ 1D CNN in this study. The convolution operation within each CNN layer can be outlined as follows:

$$\begin{aligned} \mathbf{Y} &= CONV(\mathbf{X}, \mathbf{W}, \mathbf{b}), \\ \mathbf{Y} &= BN(\mathbf{Y}), \\ \mathbf{Y} &= ReLU(\mathbf{Y}). \end{aligned} \tag{2.4}$$

The weighted input is represented by $\mathbf{X} \in \mathbb{R}^{T \times M}$, while $\mathbf{Y}$ signifies the intermediate features. Within the 1D CNN framework, $\mathbf{W}$ and $\mathbf{b}$ serve as the trainable parameters. These components collectively form the CNN block, constructed by sequentially stacking multiple CNN layers. Further insights into the optimized layers of the CNN block are provided in Sect. 2.3.

### *2.2.4 Icing Detection of Wind Turbine Blades*

In this research, we tackle the critical task of detecting icing on wind turbine blades, framing it as a binary classification challenge. Leveraging the TACNN, feature representations are extracted and subsequently processed through a global average pooling (GAP) layer and an FC network. The ultimate goal is to employ the softmax function in the output layer, generating probabilities that indicate whether the turbine blade is experiencing icing or is in a normal (non-icing) state.

Considering the feature set $\mathbf{X} \in \mathbb{R}^{K \times T}$, where $K$ denotes the number of filters in the CNN layer and $T$ represents the length of the time series, we compute the output probability $P_k \in [0, 1]$ for the icing condition through a meticulous calculation process:

$$P_k = softmax(\Phi(GAP(s))) = \frac{exp(\Psi^k \Phi(GAP(s)))}{\sum_{k=1}^{n} exp(\Psi^k \Phi(GAP(s)))}. \tag{2.5}$$

In this research, the FC network's parameters are symbolized by $\Phi$, while $\Psi$ denotes the parameters of the output layer, and $\sum_{k=1}^{n} P_k = 1, k \in [0, 1]$. These parameters, $\Phi$ and $\Psi$, undergo automatic updates and optimization during training, based on the provided training samples.

The model training employs the back-propagation algorithm for gradient descent, with cross-entropy serving as the selected loss function. To ensure efficient computation and minimize memory usage, the Adam algorithm is chosen to optimize the loss function.

## 2.3 Experiment

All experiments are implemented on a server equipped with Intel processors (64 GB) and TITAN V (12 GB). Pytorch is used for the implementation of the models. Throughout the training process, the learning rate is set to 1e-4.

### *2.3.1 Benchmark Comparison*

The proposed TACNN is positioned as a versatile solution applicable to various Time Series Classification (TSC) tasks. Its effectiveness is initially assessed across ten public benchmark datasets drawn from [28]. These datasets span diverse applications, including Face Detection, Heartbeat analysis, and PenDigits recognition, with varying numbers of output classes. For example, Face Detection involves only two classes, while Articulary Word Recognition entails 25. Dataset sizes also vary, ranging from hundreds of kilobytes to several megabytes. Classification accuracy serves as the primary evaluation metric. We compute average accuracy and the number of Wins/Ties to facilitate a comprehensive comparison with alternative approaches.

The state-of-the-art approaches used for evaluation are as follows:

- **TapNet** [23]: TapNet, a recent addition to TSC, introduces a novel approach based on prototype utilization. Leveraging a parallel architecture of Long Short-Term Memory (LSTM) and Convolutional Neural Network (CNN), TapNet autonomously extracts features and subsequently classifies them based on feature-to-prototype distances.
- **MLSTM-FCN** [20]: Comprising LSTM and Fully Convolutional Network (FCN) in parallel, MLSTM-FCN incorporates a Squeeze-and-Excitation (SE) module within FCN.
- **WEASEL-MUSE** [29]: WEASEL-MUSE is a feature-oriented model designed specifically for TSC applications.
- **ED-1NN**, **DTW-1NN-I**, **DTW-1NN-D**, **ED-1NN-N**, **DTW-1NN-I-N**, and **DTW-1NN-D-N**: These methods employ Euclidean Distance (ED) and Dynamic Time Warping (DTW) as distance metrics. I signifies individual dimension consideration in DTW, D indicates data normalization, and N denotes norm application.

The hyperparameters for TACNN are set as follows: LSTM has 64 hidden units, and the CNN consists of three layers with filter counts of {128, 256, 128}. A single FC layer is used in the icing detection network for this experiment. The best accuracy for each dataset is highlighted in bold.

TACNN demonstrates superior average accuracy compared to existing state-of-the-art methods, as shown in Table 2.1. It achieves an average accuracy of 73.1, a notable improvement over TapNet, which scores an average accuracy of 68.7. In terms of the number of victories or ties, TACNN outperforms other methods with eight wins, while TapNet and WEASEL+MUSE only achieve one win or tie each. These results suggest that TACNN performs better across most datasets, especially in datasets with limited data, such as Heartbeat and HandMovementDirection, which contain only a few hundred training samples.

**Table 2.1** Benchmark comparison in UEA multivariate time-series dataset

| Methods | AWR | AF | BM | FD | HMD | HB | NA | PD | SR | SWJ | AVG | W/T |
|---|---|---|---|---|---|---|---|---|---|---|---|---|
| TACNN | 98.3 | 46.7 | 97.5 | 62.9 | 44.6 | 75.6 | 96.1 | 98.8 | 57.2 | 53.3 | 73.1 | 8 |
| TapNet | 98.7 | 33.3 | 100.0 | 55.6 | 37.8 | 75.1 | 93.9 | 98.0 | 55.0 | 40.0 | 68.7 | 1 |
| MLSTM-FCN | 97.3 | 26.7 | 95.0 | 54.5 | 36.5 | 66.3 | 88.9 | 97.8 | 47.2 | 6.7 | 61.7 | 0 |
| WEASEL+MUSE | 99.0 | 33.3 | 100.0 | 54.5 | 36.5 | 72.7 | 87.0 | 94.8 | 46.0 | 33.3 | 65.7 | 1 |
| ED-1NN | 97.0 | 26.7 | 67.5 | 51.9 | 27.9 | 62.0 | 86.0 | 97.3 | 48.3 | 20.0 | 58.5 | 0 |
| DTW-1NN-1 | 98.0 | 26.7 | 100.0 | 51.3 | 30.6 | 65.9 | 85.0 | 93.9 | 53.3 | 33.3 | 63.8 | 1 |
| DTW-1NN-D | 98.7 | 20.0 | 97.5 | 52.9 | 23.1 | 71.7 | 88.3 | 97.7 | 53.9 | 20.0 | 62.4 | 0 |
| ED-1NN-N | 97.0 | 26.7 | 67.6 | 51.9 | 27.8 | 61.9 | 85.0 | 97.3 | 48.3 | 20.0 | 58.4 | 0 |
| DTW–1NN-I-N | 98.0 | 26.7 | 100.0 | 50.0 | 30.6 | 65.8 | 85.0 | 93.9 | 53.3 | 33.3 | 63.7 | 1 |
| DTW–1NN-D-N | 98.7 | 22.0 | 97.5 | 52.9 | 23.1 | 71.7 | 88.3 | 97.7 | 53.9 | 20.0 | 62.6 | 0 |

### 2.3.2 Dataset and Evaluation Metrics

In this study, icing data pertaining to wind turbine blades are sourced from Goldwind Inc., a prominent global wind turbine manufacturer. Operational data from three wind turbines in Inner Mongolia, China, are logged by the SCADA system. The recorded running times for these turbines are 305.77, 695.59, and 329.28 h, respectively.

Intermittent stoppages occur in the first two turbines, while continuous logging is available for the third turbine. The raw sensor data collected by the SCADA system encompass numerous dimensions. However, engineers have narrowed down the parameters to 26, focusing specifically on icing-related aspects and removing irrelevant parameters based on domain expertise, as shown in Table 2.2. Moreover, engineers have provided labels indicating the occurrence range of icing. The labeled data exhibit significant class imbalance, necessitating the application of imbalanced data processing techniques detailed in Sect. 2.2.2. For offline training and testing, data from the two turbines with intermittent sensor data are combined. Approximately 80% of the data is allocated for training, with the remaining 20% reserved for testing. Conversely, data from the continuously logged turbine are utilized for online estimation of icing probability.

The evaluation of models employs several metrics, including **Precision**, **Recall**, **F1**, and **Matthews correlation coefficient** (**MCC**). The definitions of these metrics are outlined as follows:

$$Precision = \frac{TP}{TP + FP}, \quad Recall = \frac{TP}{TP + FN}, \tag{2.6}$$

$$F1 = \frac{2 \times Precision \times Recall}{Precision + Recall}, \tag{2.7}$$

$$MCC = \frac{TP \times TN - FP \times FN}{\sqrt{(TP + FP)(TP + FN)(TN + FP)(TN + FN)}}, \tag{2.8}$$

**Table 2.2** Wind Turbine SCADA data parameters

| Index | Monitored variable | Index | Monitored variable |
|---|---|---|---|
| 1 | Wind velocity | 14 | Pitch motor 1 temperature |
| 2 | Generator rotational speed | 15 | Pitch motor 2 temperature |
| 3 | Active electrical power output | 16 | Pitch motor 3 temperature |
| 4 | Wind direction (Compass Heading) | 17 | Horizontal acceleration (Turbine Structure) |
| 5 | 25-Second averaged wind Direction Angle | 18 | Vertical acceleration (Turbine Structure) |
| 6 | Yaw system position | 19 | Ambient temperature |
| 7 | Yawing Rate | 20 | Nacelle internal temperature |
| 8 | Pitch Angle–Blade 1 | 21 | Pitch 1 switching element temperature |
| 9 | Pitch Angle–Blade 2 | 22 | Pitch 2 switching element temperature |
| 10 | Pitch Angle–Blade 3 | 23 | Pitch 3 switching element temperature |
| 11 | Pitch actuator 1 speed | 24 | Pitch 1 Charger DC power consumption |
| 12 | Pitch actuator 2 speed | 25 | Pitch 2 charger DC power consumption |
| 13 | Pitch actuator 3 speed | 26 | Pitch 3 charger DC power consumption |

where $TP$, $FP$, $FN$, and $TN$ represent true positive, false positive, false negative, and true negative, respectively.

### 2.3.3 Baseline Comparison

To thoroughly assess the effectiveness of our proposed approach in detecting icing on wind turbine blades, we conduct a comprehensive comparison with ten baseline networks. Each baseline network offers unique architectures and characteristics, providing valuable insights into the performance landscape. Here, we delineate the specifics of these baseline networks:

- **MultiLayer Perceptron (MLP)**: This architecture features three fully connected (FC) layers, each equipped with $\{500\}$ hidden nodes. To combat overfitting, dropout layers are strategically placed between FC layers.
- **Long Short-Term Memory (LSTM)**: Renowned for its adeptness in modeling time-series data, LSTM is employed with a single layer. We explore varying numbers of hidden units from $\{8, 16, 32, 64\}$, selecting the optimal model for comparison.

- **Gated Recurrent Unit (GRU)**: A lighter variant of LSTM, GRU boasts comparable performance. We experiment with different hidden unit numbers from {8, 16, 32, 64}, selecting the top-performing model.
- **CNN**: Leveraging a convolutional neural network (CNN) layer with parameters and structure detailed in Sect. 2.2.3.
- **FCN**: Recognized for its competitive edge in Time Series Classification (TSC), FCN integrates three convolutional layers with filter sizes of {128, 256, 128}.
- **ResNet**: A variant of CNN widely acclaimed in TSC, adapted from [19].
- **WaveletFCNN**: Tailored specifically for blade icing detection in wind turbines, WaveletFCNN synergizes wavelet transformation and fully convolutional neural network (FCNN) methodologies, as detailed in [6].
- **MSCNN**: A cutting-edge multi-scale deep CNN tailored for fault diagnosis in wind turbine gearboxes. We meticulously re-implement the model based on the specifications outlined in [8], considering three distinct scales.

The **MLSTM-FCN** and **TapNet** models, detailed in Sect. 2.3.1, serve as benchmarks for comparison. Our proposed model operates as an end-to-end deep learning architecture. Consequently, only these two deep learning models, as described in Sect. 2.3.1, are employed for comparative analysis in this section.The hyperparameters of the TACNN remain consistent with those outlined in Sect. 2.3.1, with the exception of the icing detection network, which now incorporates three fully connected (FC) layers. Furthermore, the window size (T) for the sensor data is fixed at 128 time steps, roughly equivalent to 15 minutes of data.

As presented in Table 2.3, our proposed model exhibits superior performance compared to other baseline methods. Notably, it achieves remarkable improvements of approximately 16.3% and 6.83% in Matthews correlation coefficient (MCC) and F1-score, respectively, over the best-performing ResNet model. Furthermore, compared to the weakest-performing model, our model demonstrates substantial enhancements

**Table 2.3** Performance of baseline comparison on SCADA data

| Models | Recall | Precision | F1 | MCC |
|---|---|---|---|---|
| MLP | 60.7 | 71.8 | 71.4 | 42.0 |
| LSTM | 65.9 | 86.3 | 79.9 | 60.1 |
| GRU | 59.4 | 85.1 | 77.0 | 54.4 |
| CNN | 55.6 | 85.1 | 75.6 | 52.0 |
| FCN | 73.0 | 87.9 | 83.3 | 66.5 |
| ResNet | 88.5 | 77.8 | 83.4 | 67.4 |
| TapNet | 71.0 | 77.6 | 77.7 | 54.7 |
| MLSTM-FCN | 65.7 | 83.3 | 78.6 | 57.0 |
| WaveletFCNN | 44.5 | **92.0** | 73.2 | 49.8 |
| MSCNN | 68.7 | 83.7 | 79.8 | 59.4 |
| TACNN | **92.2** | 85.0 | **89.1** | **78.4** |

of 24.8% and 86.7% in F1-score and MCC, respectively.LSTM and GRU models, leveraging their capability to capture periodic features, outperform MLP marginally. Impressively, our proposed model stands as the sole method for achieving a recall accuracy exceeding 90%. However, WaveletFCNN surpasses our proposed TACNN in precision accuracy. ResNet and FCN achieve comparable levels of accuracy across F1-score and MCC metrics. Surprisingly, TapNet and MLSTM-FCN fail to attain the desired accuracy levels, possibly due to the intricate hyperparameter optimization required for improved performance. Additionally, their design for multivariate time-series classification (TSC) problems may not be well-suited for the nuances of wind turbine blade icing detection.WaveletFCNN and MSCNN, tailored for fault diagnosis in wind turbines, do not meet expectations. Our TACNN outperforms both models by significant margins, demonstrating improvements of 57.4% and 32.0% in MCC, and 21.7% and 11.7% in F1-score, respectively. Notably, WaveletFCNN and MSCNN utilize multi-scale features, employing wavelets and multi-scale entropy, respectively. However, their underperformance could stem from the specific variants selected for comparison, and the limitations of MSCNN in extracting discriminative features for blade icing from sensor data. In conclusion, our proposed TACNN emerges as a robust solution for wind turbine blade icing detection, offering substantial performance gains over existing methods and holding promise for real-world applications in renewable energy.

### *2.3.4 Comparison with Other Attention Mechanisms*

To further demonstrate the effectiveness of our proposed approach, we compare it with three commonly used attention mechanisms using the icing datasets, as shown in Table 2.4. These attention mechanisms are integrated after the CNN layer, with consistent CNN structures and parameters resembling those of the TACNN. Each CNN layer is augmented with a specific attention module, namely: **SE** [30]: A well-known attention module designed for CNN architectures. **CBAM** [25]: Comprising both channel and spatial attention modules for comprehensive feature refinement. **GC** [26]: A lightweight attention mechanism adept at capturing global context, enhancing feature learning capabilities.

**Table 2.4** Comparison with other attention modules

| Models | Recall | Precision | F1 | MCC |
|---|---|---|---|---|
| SE | 72.7 | 84.8 | 81.8 | 63.3 |
| CBAM | 86.3 | 78.0 | 82.9 | 66.0 |
| GC | **99.6** | 74.6 | 84.5 | 72.9 |
| TACNN | 92.2 | **85.0** | **89.1** | **78.4** |

Table 2.4 clearly demonstrates that our proposed TACNN outperforms other attention modules in terms of Matthews correlation coefficient (MCC), F1-score, and Precision. Notably, GC achieves the second-best performance among the three attention modules, attributed to its superior feature learning capabilities. Specifically, our TACNN showcases improvements of 23.9%, 18.8%, and 7.54% over SE, CBAM, and GC, respectively, in terms of MCC. Regarding F1-score, SE and CBAM demonstrate similar classification accuracy, with a notable 5.44% improvement observed between TACNN and GC. Interestingly, GC achieves nearly 100% accuracy in Recall, surpassing our proposed TACNN, owing to its ability to explore feature relationships in a global context.

The primary distinction between our proposed TACNN and widely used attention modules lies in their application within the model architecture. Our proposed attention mechanism operates directly on raw sensor data, preceding the CNN layers, whereas conventional attention mechanisms re-weight features extracted by the CNN. Experimental findings demonstrate the efficacy of our proposed TACNN in enhancing performance. This can be attributed to the model's ability to assess the importance of sensors and time steps directly, a capability not inherent in conventional attention mechanisms designed for Time Series Classification (TSC) tasks. Unlike conventional attention mechanisms that rely on CNN-extracted features, our method processes sensor data prior to CNN input. This approach allows for direct evaluation of sensor and time step importance, contributing to improved performance. In essence, the superiority and significance of our proposed attention mechanism stem from its reinforcement of the icing status learning mechanism and exploitation of discriminative feature learning. In summary, our proposed TACNN offers distinct advantages by leveraging attention directly on raw sensor data.

### *2.3.5 Ablation Study and Sensitivity Analysis*

An ablation study was conducted to assess the significance of the proposed Temporal Attention (TA) module within the TACNN architecture. For this purpose, the TA module was excluded, resulting in a variant termed **TACNN_TA**. Comparative analysis of TACNN and TACNN_TA was performed across four distinct datasets, employing window sizes of 32, 64, 128, and 256. Figure 2.3 depicts the performance metrics, namely F1-score and Matthews Correlation Coefficient (MCC), for both TACNN and TACNN_TA across these varying window sizes. The results reveal several key findings: The inclusion of the TA module indeed enhances the performance of the conventional CNN architecture. Notably, the most significant decrease in accuracy occurs when the window size is set to 128. It is observed that the performance of TACNN exhibits a pattern of initial improvement followed by a subsequent decline as the window size varies. Specifically, TACNN achieves optimal performance at a window size of 128, after which its performance gradually diminishes. In summary, the findings suggest that the proposed TACNN architecture is highly sensitive to the temporal length of the input time-series data.

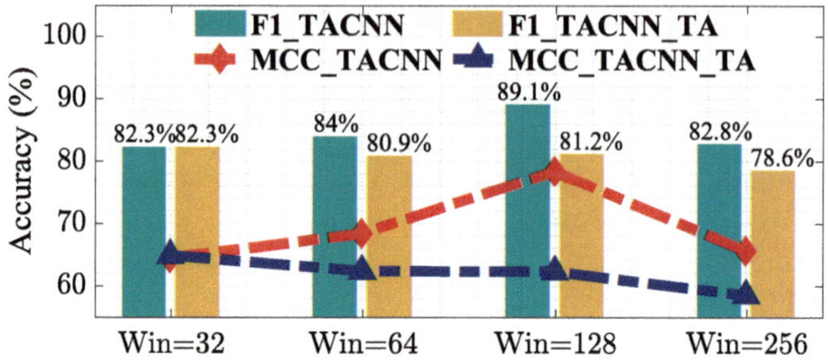


**Fig. 2.3** Ablation analysis

To investigate the influential factors affecting TACNN and to optimize its architecture, a sensitivity analysis was conducted. The findings are depicted in Figs. 2.4, 2.5, and 2.6. Exploring the impact of the number of CNN layers in TACNN, four distinct network configurations were devised. These configurations correspond to varying numbers of filters: {128}, {128, 256}, {128, 256, 128}, and {128, 256, 128, 128}, respectively.To assess the influence of filter size on TACNN, the number of layers was fixed at three while altering the filter sizes: {32, 64, 128, 256}. Furthermore, the importance of hidden units within the TA module was examined. This analysis involved setting the CNN layers to three, with a predetermined filter size of {128, 256, 128}, while varying the number of hidden units across a range from 16 to 128. This comprehensive sensitivity analysis provides insights into key architectural parameters affecting the performance of TACNN, facilitating the optimization of its structure.

Figure 2.4 reveals that the highest Precision, F1, and MCC scores are achieved with three CNN layers, contrasting with the suboptimal performance observed with four layers instead of one. This disparity can be attributed to the number of filters, suggesting potential improvements through filter optimization. Figure 2.5 further

**Fig. 2.4** SA of CNN layers

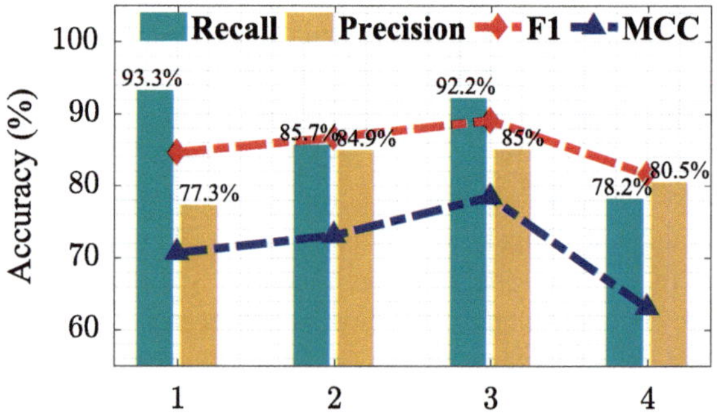

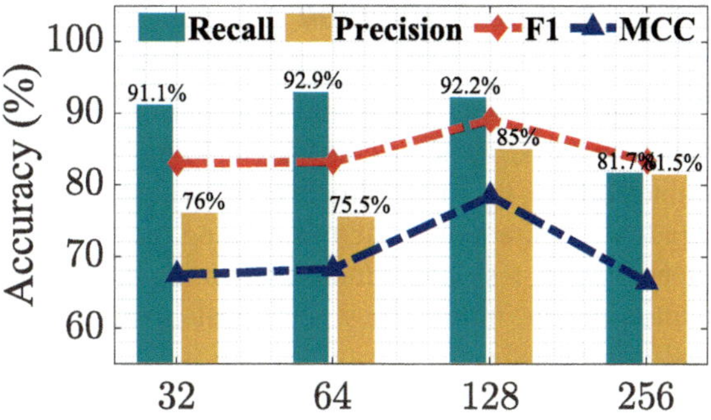


**Fig. 2.5** SA of filter size in CNN

supports this notion, indicating peak Precision, F1, and MCC scores when the CNN filter size is set to 128. Notably, there's a consistent trend of increasing accuracy, particularly in F1 and MCC, as the filter size increases from 32 to 128, albeit a decline when the size is 256. In addition, Fig. 2.6 demonstrates that optimal performance is achieved with 64 hidden units within the TA module. A discernible trend emerges, showcasing performance enhancement as the number of hidden units rises from 16 to 64. However, a notable decline in performance occurs beyond 64 units, even underperforming compared to 32 units. This analysis underscores the critical role of architectural parameters, including the number of CNN layers, filter size, and hidden units within the TA module, in shaping the performance of TACNN.

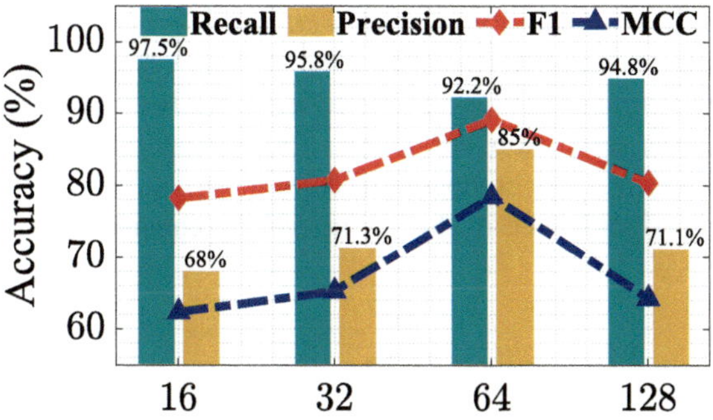


**Fig. 2.6** SA of hidden units in TA

### 2.3.6 Online Estimation

A novel framework for real-time detection of icing conditions on wind turbine blades within wind farms is introduced, employing an online estimation scheme. Initially trained offline, the model processes historical sensor data by segmenting it into fixed window sizes, a pivotal step in training the icing detector. Transitioning to online identification, a sliding window, mirroring the length of the segments, dynamically adjusts with the influx of new sensor data, enabling the model to continuously furnish real-time probabilities regarding icing and normal conditions. To fortify the reliability of real-time estimations and to circumvent inadvertent errors, the majority vote algorithm is harnessed for robust decision-making, a method whose efficacy has been corroborated [6]. As expounded in Sect. 2.2.2, this adaptive approach, coupled with the uninterrupted logging of sensor data, empowers the system for seamless online icing detection, enhancing operational safety and efficiency in wind farm operations. Throughout the experimental phase, the hyperparameters remain consistent with established configurations. The model's predictive capabilities, delivering insights approximately every 4 minutes, are tested over a comprehensive duration spanning around 23 h.

Figure 2.7 showcases the online estimation performance of TACNN and TACNN_TA, providing further insight into the efficacy of the proposed Temporal Attention (TA) module. The magenta background highlights icing periods, designated by domain experts, while a green dashed line represents the classical threshold (0.5) for icing detection. During icing periods, TACNN achieves 100% accuracy, whereas TACNN_TA exhibits around 50% accuracy. In normal conditions, TACNN's output probability exhibits a gradual increase, whereas TACNN_TA's output fluctuates abruptly. This difference underscores the role of the TA module in capturing temporal dependencies between successive time steps. Given the gradual nature of ice accumulation, our model affords engineers more time for proactive icing prevention measures compared to conventional CNN approaches. The results underscore the significance of the TA module in enhancing TACNN's predictive capabilities,

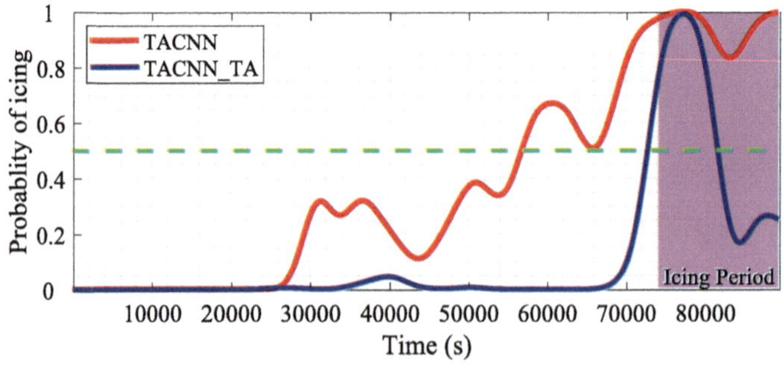


**Fig. 2.7** Results of online estimation of TACNN and TACNN_TA

particularly in scenarios where early detection and proactive intervention are critical for mitigating icing-related risks.

In the realm of recorded SCADA data, the operational states are simplified into two categories: normal and icing. However, the process of ice accumulation on turbine blades is not binary; it unfolds gradually with varying degrees of intensity. This intricate nature of icing poses a significant challenge for deep learning models tasked with accurately discerning the severity of icing on wind turbine blades. Achieving early detection of icing events is paramount for effective operational management.One promising avenue to address this challenge involves fine-tuning the sensitivity of the model for early icing detection. This is accomplished by interpreting the output probability defined in Eq. 2.5 and adjusting it using different thresholds. For instance, by setting a threshold of 0.3, the proposed TACNN can swiftly identify icing events at their nascent stages, albeit with occasional misclassifications during normal operating conditions. Conversely, employing a higher threshold, such as 0.75, ensures that the TACNN maintains competitive performance levels across both normal and icing periods. In practical applications, determining the optimal threshold necessitates a thorough examination of the relationship between the chosen threshold and the observed amount of icing.

## 2.4 Conclusion

This research introduces a novel deep learning architecture designed specifically for identifying icing conditions on wind turbine blades. By integrating traditional CNN with a TA module, the model aims to understand the significance of sensors and time steps while automatically extracting discriminative features from raw temporal sensor data. The effectiveness of the proposed TACNN in addressing the challenges highlighted in Sect. 2.1 has been successfully demonstrated. Moreover, the TACNN demonstrates its versatility as a general solution for Time Series Classification (TSC), as validated across ten publicly available benchmark datasets. Applied to icing datasets obtained from three wind turbines operated by a leading wind power company, the TACNN outperforms ten state-of-the-art baseline networks, exhibiting significant accuracy advantages. Additionally, in comparison to three widely adopted attention mechanisms, the proposed model yields competitive results. Its generalizability and practical utility are further validated through online estimation testing, confirming its efficacy in real-world applications.

Two fundamental assumptions underlie the proposed model: (1) Consistency between the feature space and label space of both training and testing samples, necessitating identical window sizes and sampling frequencies. (2) Similarity in the distribution of training and testing samples. However, several limitations affect the efficacy of TACNN. Firstly, TACNN lacks the capability to discern the severity of icing on wind turbine blades. Addressing this limitation may entail structural modifications to the model, enhancing its capacity for robust feature learning in assessing icing conditions. Secondly, TACNN's reliance on individualized training for each wind turbine restricts its generalizability. Exploring techniques like transfer learning could facilitate the adaptation of TACNN to diverse wind turbine sizes and types, thereby improving its applicability and versatility. Thirdly, blindly applying TACNN to vibrating systems poses inherent risks. The model's alignment with system dynamics cannot be guaranteed, hindering extrapolation efforts. Combining TACNN with mathematical modeling methods or Gaussian processes [31] may offer a viable solution, enabling the capture of system dynamics in an interpretable manner. Furthermore, future research should focus on hyperparameter optimization to refine the network structure and enhance model performance.

## References

1. Ren Z, Verma AS, Li Y, Teuwen JJE, Jiang Z (2021) Offshore wind turbine operations and maintenance: a state-of-the-art review. Renew Sustain Energy Rev 144:110886
2. Wang Y, Yuanming X, Lei Y (2018) An effect assessment and prediction method of ultrasonic de-icing for composite wind turbine blades. Renew Energy 118:1015–1023
3. Wei K, Yang Y, Zuo H, Zhong D (2020) A review on ice detection technology and ice elimination technology for wind turbine. Wind Energy 23(3):433–457
4. Madi E, Pope K, Huang W, Iqbal T (2019) A review of integrating ice detection and mitigation for wind turbine blades. Renew Sustain Energy Rev 103:269–281
5. Zhang M, Jing Yu, Chen R, Liu Q, Liu J, Song D, Liu P, Gao L, Wang J (2019) Highly transparent and robust slippery lubricant-infused porous surfaces with anti-icing and anti-fouling performances. J Alloy Compd 803:51–60
6. Yuan B, Wang C, Jiang F, Long M, Yu PS, Liu Y (2019) Waveletfcnn: a deep time series classification model for wind turbine blade icing detection. arXiv: 1902.05625
7. Shu L, Qiu G, Qin H, Jiang X, McClure G, Yang H (2018) Numerical and field experimental investigation of wind turbine dynamic de-icing process. J Wind Eng Ind Aerodyn 175:90–99
8. Jiang G, He H, Yan J, Xie P (2018) Multiscale convolutional neural networks for fault diagnosis of wind turbine gearbox. IEEE Trans Ind Electron 66(4):3196–3207
9. Liu Y, Cheng H, Kong X, Wang Q, Cui H (2019) Intelligent wind turbine blade icing detection using supervisory control and data acquisition data and ensemble deep learning. Energy Sci Eng 7(6):2633–2645
10. Jiménez AA, García Márquez FP, Moraleda VB, Muñoz CQG (2019) Linear and nonlinear features and machine learning for wind turbine blade ice detection and diagnosis. Renew Energy 132:1034–1048
11. Zhang L, Liu K, Wang Y, Omariba ZB (2018) Ice detection model of wind turbine blades based on random forest classifier. Energies 11(10):2548
12. Shinan C, Mengyao L, Wu H, James T (2016) Aircraft ice accretion prediction using neural network and wavelet packet transform. Int J Aircraft Eng Aerosp Technol
13. Ogretim E, Huebsch W, Shinn A (2006) Aircraft ice accretion prediction based on neural networks. J Aircraft 43(1):233–240
14. Hasan Huseyin Yildirim and Mehmet Yavuz (2019) Evaluation of wind energy investment with artificial neural networks. Int J Optim Control: Theor Appl (IJOCTA) 9(2):142–147
15. Orsenigo C, Vercellis C (2010) Combining discrete SVM and fixed cardinality warping distances for multivariate time series classification. Pattern Recogn 43(11):3787–3794
16. Baydogan MG, Runger G, Tuv E (2013) A bag-of-features framework to classify time series. IEEE Trans Pattern Anal Mach Intell 35(11):2796–2802
17. Schäfer P (2016) Scalable time series classification. Data Mining Knowl Disc 30(5):1273–1298
18. Pei W, Dibeklioğlu H, Tax DMJ, van der Maaten L (2018) Multivariate time-series classification using the hidden-unit logistic model. IEEE Trans Neural Netw Learn Syst 29(4):920–931

19. Wang Z, Yan W, Oates T (2017) Time series classification from scratch with deep neural networks: a strong baseline. In: 2017 international joint conference on neural networks (IJCNN). IEEE, pp 1578–1585
20. Karim F, Majumdar S, Darabi H, Harford S (2019) Multivariate LSTM-FCNS for time series classification. Neural Netw 116:237–245
21. Cheng X, Li G, Skulstad R, Chen S, Hildre HP, Zhang H (2019) Modeling and analysis of motion data from dynamically positioned vessels for sea state estimation. In: 2019 international conference on robotics and automation (ICRA). IEEE, pp 6644–6650
22. Cheng X, Li G, Ellefsen AL, Chen S, Hildre HP, Zhang H (2020) A novel densely connected convolutional neural network for sea state estimation using ship motion data. IEEE Trans Instrum Measurem 69(9):5984–5993
23. Zhang X, Gao Y, Lin J, Lu C-T (2020) Tapnet: multivariate time series classification with attentional prototypical network. In: Proceedings of the AAAI conference on artificial intelligence
24. Shoja S (2016) Guided wave propagation in composite structures. Master's thesis, Chalmers University of Technology
25. Woo S, Park J, Lee J-Y, Kweon IS (2018) Cbam: convolutional block attention module. In: Proceedings of the European conference on computer vision (ECCV), pp 3–19
26. Cao Y, Xu J, Lin S, Wei F, Hu H (2019) Gcnet: non-local networks meet squeeze-excitation networks and beyond. In: Proceedings of the IEEE international conference on computer vision workshops, pp 0–0
27. Qin Y, Song D, Chen H, Cheng W, Jiang G, Cottrell G (2017) A dual-stage attention-based recurrent neural network for time series prediction. IJCAI
28. Bagnall A, Dau HA, Lines J, Flynn M, Large J, Bostrom A, Southam P, Keogh E (2018) The UEA multivariate time series classification archive (2018). arXiv:1811.00075
29. Schäfer P, Leser U (2017) Multivariate time series classification with weasel+ muse. In: Proceedings of ACM conference, pp 0–0
30. Hu J, Shen L, Sun G (2018) Squeeze-and-excitation networks. In: Proceedings of the IEEE conference on computer vision and pattern recognition, pp 7132–7141
31. Weijtjens W, Avendaño-Valencia LD, Devriendt C, Chatzi E (2018) Cost-effective vibration based detection of wind turbine blade icing from sensors mounted on the tower. In: 9th European workshop on structural health monitoring, EWSHM 2018

# Chapter 3
# Wavelet-Enhanced Temporal Convolutional Neural Network for Blade Icing Detection

## 3.1 Introduction

Renewable energy resources have garnered considerable interest amid the global energy crisis and the escalating need for sustainable energy solutions [1]. Wind power, distinguished by its broad availability, economic viability, and mature technology for commercial deployment, currently represents one of the most rapidly expanding segments within the renewable energy sector globally [2]. However, blade icing severely restricts the operational efficiency of wind turbines in frigid climates. Wind farms are frequently situated in elevated mountainous regions characterized by low temperatures and high humidity, rendering wind turbine blades particularly susceptible to icing, especially during winter periods. Ice accretion on blades not only diminishes power generation, potentially causing up to 30% annual energy production losses in extreme cases, but also introduces safety hazards in the vicinity of wind power installations [3]. Consequently, the detection of icing on wind turbine blades has become an area of increasing focus. To mitigate losses attributable to blade icing, substantial research efforts have been directed towards icing condition identification.

Conventional methodologies for wind turbine blade icing detection can be broadly categorized into direct and indirect approaches. Direct techniques rely on the physical attributes of ice, whereas indirect methods depend on measuring parameters influenced by ice accumulation [4]. Direct detection methods primarily encompass ultrasonic damping [5], resonant frequency analysis [6], optical measurement technologies [7], and similar techniques. Indirect detection approaches include comparing output power [3], utilizing heated and unheated anemometers [8], monitoring dew point and temperature [6], among others. Conventional methods offer the advantage of highly dependable measurements, such as ice thickness and presence [3, 9]. However, they are notably disadvantaged by potential structural impacts on blades, substantial costs, extensive personnel requirements, limited accuracy in icing detec-



X. Cheng et al., *Computational Methods for Blade Icing Detection of Wind Turbines*, Engineering Applications of Computational Methods 24,
https://doi.org/10.1007/978-981-96-6763-5_3

tion, particularly for minor ice accumulations, and insensitivity to small quantities of ice [10].

To circumvent the limitations of conventional techniques, recent research has increasingly focused on blade icing detection methods leveraging wind turbine monitoring signals, owing to their reduced cost and minimal mechanical modifications. These methods are divided into model-driven and data-driven categories. Model-driven methods utilize domain-specific knowledge to construct mathematical models that represent the correlation between monitoring signals and icing. Corradini et al. proposed a mathematical model for wind turbine blade icing monitoring based on an observer to assess rotor angular speed error exceeding a predetermined threshold [11]. Shi presented a model-based blade icing detection technique that obviates the need for supplementary measurements and is applicable to various blade aerodynamic alterations, not solely ice formation [12]. Hu et al. developed an icing model to enhance understanding of blade icing conditions, validated through numerical simulation of ice accretion on rotating NREL Phase VI blades [13]. However, formulating a precise model for wind turbine blade icing estimation is challenging, as model-driven methods are heavily contingent on prior domain knowledge and certain assumptions. Furthermore, external resources, such as wind tunnels, may be required to adjust model parameters.

Conversely, data-driven methods employ shallow machine learning and deep learning algorithms for wind turbine blade icing detection. Shallow machine learning techniques, such as linear discriminant analysis, were used to identify power and wind speed distribution features, followed by multiple random forest classifiers for icing detection in [14]. Additionally, Georgios utilized k-Nearest Neighbors (kNN) to conduct icing failure detection based on the relationship between wind speed and power [15]. Support vector machines (SVMs) have also been applied to wind turbine icing detection in [16]. Yi et al. introduced an imbalanced classification model for wind turbine blade icing detection, presenting a novel minority clustering SMOTE (synthetic minority oversampling technique) method to address data imbalance in collected sensor data [17]. While these shallow machine learning methods do not depend on exact mathematical models, they necessitate considerable manual effort in feature engineering [18].

Deep learning methodologies eliminate the need for manual feature extraction [19]. They can autonomously learn high-level feature representations from monitoring sensor data, demonstrating the significant potential for blade icing detection in wind turbines. Chen et al. proposed a deep neural network to learn distinguishing features for model construction, explicitly addressing data imbalance in blade icing detection and employing a triplet loss function to preserve intra-class and inter-class information [20]. Liu et al. introduced a deep neural network combining deep learning and ensemble learning techniques to improve model accuracy and generalization capability [21]. Cheng et al. developed a temporal attention convolutional neural network for blade icing detection [22]. Yuan et al. [23] investigated wind turbine blade icing detection by integrating wavelet transformation with fully convolutional neural networks (FCNNs). However, the waveletFCNN approach did not account for temporal dependencies, which may limit its suitability in real-world applications. Despite the successful application of deep learning in diverse fields such as computer vision and natural language processing, its in-depth application to wind turbine blade icing detection remains underexplored. Several challenges persist in employing deep learning-based methods for detecting icing on wind turbine blades:

1. **Data Aspect**: Wind turbines predominantly operate in normal states, with icing events being comparatively infrequent. This results in a significant class imbalance in supervisory control and data acquisition (SCADA) system monitoring data, where normal state data far outweighs abnormal (icing) state data. Training icing detection models on such imbalanced datasets can lead to biased identification of blade icing.
2. **Incomplete Information**: Data acquired from SCADA systems are inevitably prone to information loss due to sensor malfunctions. Common strategies for handling missing data include imputation, mean substitution, etc. However, relying solely on data completion algorithms for incomplete data can lead to accumulated errors in detection tasks.
3. **Model Aspect**: Wind turbines function within complex engineering systems under variable operational conditions, leading to monitoring signals with complex and diverse features. Consequently, extracting meaningful features from intricate multivariate signals using deep learning methods to accurately detect blade icing presents a considerable challenge.
4. **Application Aspect**: Data-driven models are typically trained offline and subsequently deployed for real-time estimation. Although high accuracy can be achieved during offline training and testing, the more complex real-world environment can result in elevated error or false alarm rates when deploying these offline-trained models in practical application scenarios. Therefore, robust methodologies are needed for real-world implementations.

To overcome these challenges, we propose a multilevel convolutional recurrent neural network (MCRNN). Specifically, for the data aspect, we perform data analysis and preprocessing to minimize the impact of noise and redundant information. To address severe data imbalance, we investigate two solutions: data resampling algorithms and class-balanced loss functions. For incomplete information, we employ different interpolation techniques based on the duration of missing value intervals, including linear interpolation and direct interpolation using data from the same period in different years. For the model aspect, we initially apply multilevel discrete wavelet decomposition (MDWD) to the raw data to obtain wavelet detail coefficients that capture signal variations across multiple scales, revealing information in both time and frequency domains. Subsequently, a parallel structure combining an LSTM branch and a CNN branch is established at each level to learn from the correlations within multivariate signals. Finally, to ensure high availability in real-world applications, we introduce a multi-step accumulation strategy to reduce error estimation and false alarm rates during model deployment.

The primary contributions of this study are:

1. We introduce a novel deep neural network, MCRNN, for wind turbine blade icing detection. MDWD is initially employed to extract multilevel features. A parallel network architecture is then utilized at each level to capture complex correlations in both frequency and temporal domains. We explore data resampling and loss function-based methods to mitigate monitoring data imbalance. A multi-step accumulation strategy is proposed to effectively decrease the false alarm rate in real-time detection.
2. The performance of the proposed MCRNN is assessed using real-world datasets from two wind turbines. Through comparisons with baseline models, MCRNN demonstrates superior effectiveness and performance. Further evaluations and comparisons are conducted with several model variants. Sensitivity analysis of MCRNN determines optimal hyperparameters. Real-time prediction validates the practical applicability of the method.

The remainder of this paper is structured as follows: Sect. 3.2 details the overall architecture of the proposed MCRNN. Section 3.3 presents experimental validation of the effectiveness and superiority of the proposed model. Section 3.4 concludes the paper.

## 3.2 Methodology

This section elucidates the comprehensive architecture of the proposed multilevel convolutional recurrent neural network for icing detection. We begin by introducing the overall workflow, followed by a detailed description of each component.

### *3.2.1 Overview*

As depicted in Fig. 3.1, the general framework of our multilevel convolutional recurrent neural network for icing detection comprises data collection and processing, feature extraction and knowledge acquisition using MCRNN, and icing detection. Sensor data collected is inevitably corrupted due to various factors, necessitating data cleaning to minimize the influence of noise. Furthermore, to address the data challenges outlined in Sect. 3.1, imbalanced data are analyzed and processed. The refined sensor data is then fed into the MCRNN for feature extraction and knowledge learning, as detailed in Sect. 3.2.3. Finally, features extracted by MCRNN are input into a custom classifier for icing detection, and optimized models are utilized for real-time icing detection.

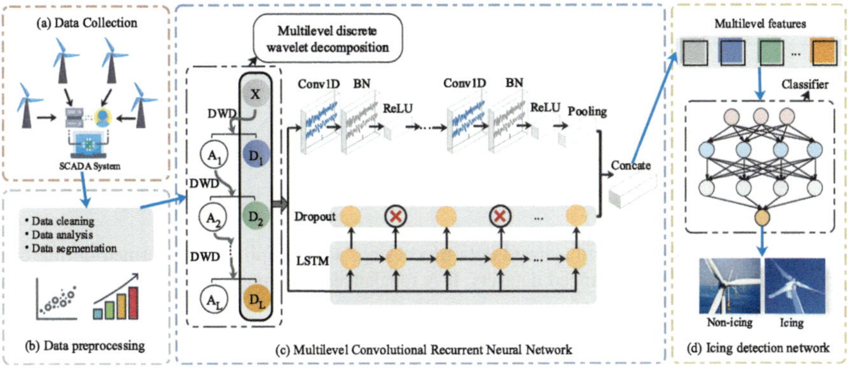


**Fig. 3.1** Overall structure of the proposed method. **a** The SCADA system collects wind turbine operating data. **b** Analysis and preprocessing of the raw data. **c** The processed datasets are input to the proposed MCRNN model for feature extraction and knowledge learning. $X$ is the input sensor data, $A_l$ and $D_l$ represent the approximate coefficient and detail coefficient of each level in multilevel discrete wavelet decomposition. BN and ReLU indicate the batch normalization layer and the ReLU layer, respectively. **d** Feature classification and icing detection

### *3.2.2 Data Processing*

The data employed in this study are sourced from the SCADA system, utilizing only existing sensors. The dataset primarily encompasses information regarding wind turbine component status and both internal and external environmental conditions. Data processing in this work involves four stages: data labeling, visualization, imbalanced data handling, and data splitting and normalization.

1. **Data Labeling**. Raw data are annotated by experienced engineers. Due to the erratic operation of wind turbines, ambiguous intervals are excluded. These uncertain periods are challenging to classify definitively as either normal or icing states, even for seasoned professionals.
2. **Missing Value Processing**. The issue of missing data in monitoring signals is prevalent due to factors such as sensor degradation and human error. For isolated or minor data gaps, linear interpolation is used to estimate missing values. However, when data loss extends over several hours or even a day, the operational state of the wind turbine may have undergone multiple transitions, rendering interpolation algorithms prone to significant errors. In such instances, data from the same time frame in different years are employed for interpolation.
3. **Statistics**. Statistical methods are valuable for understanding sensor data and comprehensively characterizing features of labeled data. Statistics offers two key advantages: (i) Identification of redundant information that minimally contributes to icing detection. Such superfluous data are discarded as they can impair feature representation. (ii) Visualization of signal correlations. A high correlation between signals can lead to severe model overfitting. To mitigate this, the average value

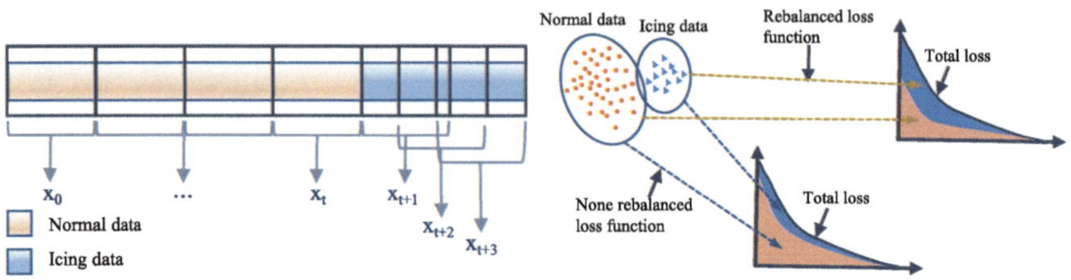


**Fig. 3.2** Two solutions for imbalanced data processing. **a** Resampling is performed during data segmentation to generate balanced datasets. **b** A rebalanced loss function is used to optimize the model

of highly correlated variables is used as a surrogate variable for icing detection in this study, replacing the original correlated variables.

4. **Imbalanced Data Processing**. Class imbalance is a widespread issue in real-world datasets [24]. SCADA system data often exhibit significant imbalance because wind turbines typically operate under normal conditions for the majority of the time, with icing events being infrequent. If not properly addressed, model training with imbalanced data can bias prediction outcomes. To tackle this, we explore two strategies: data resampling and class-balanced loss functions. As shown in Fig. 3.2a, a resampling method is used to generate balanced datasets. For normal data, samples are created using a non-overlapping sliding window. For icing data, to achieve an equal number of samples, a sliding window with overlap is applied. For instance, in Fig. 3.2a, we obtain four normal data samples and four icing data samples using our resampling approach, compared to only two icing samples without overlap. As illustrated in Fig. 3.2b, a class-rebalanced loss function is employed to optimize the model by assigning greater weights to the minority class (icing data).
5. **Data Splitting and Normalization**. SCADA system output data consist of discrete sequences because the system samples signals from various sensors approximately every 7 s. To ensure each MCRNN input sample spans a meaningful time interval, original data are segmented using a fixed step size. Furthermore, data normalization is crucial for preprocessing. It eliminates unit dependencies and converts data to dimensionless pure values, enabling simultaneous analysis of indicators with differing units or magnitudes.

### *3.2.3 Multilevel Convolutional Recurrent Neural Network*

The proposed multilevel convolutional recurrent neural network is composed of three modules: discrete wavelet decomposition, an LSTM branch, and a CNN branch. As illustrated in Fig. 3.1c, raw monitoring signals are initially transformed using MDWD to extract multilevel features. Subsequently, both the original signals and the discrete wavelet decomposition coefficients at each level are processed in parallel by the LSTM and CNN branches. The LSTM branch is designed for temporal feature learning, while the CNN branch is intended to capture spatial information.

#### 3.2.3.1 Multilevel Discrete Wavelet Decomposition

Wind turbines operate in complex engineering environments with fluctuating conditions, leading to monitoring signals acquired by SCADA systems that possess intricate and diverse characteristics. Wavelet decomposition is a well-established technique for time-series analysis due to its capability to extract features from both time and frequency domains [25]. Transforming signals via MDWD is advantageous for obtaining more actionable information.

MDWD decomposes a time series into multilevel sub-series ordered by frequency from high to low. These frequencies are valuable for feature learning in the frequency domain. Let $x_c = [x_0, \ldots, x_n, \ldots, x_{N-1}]$ represent a sequence segment for each channel, where $c = 1, 2, \ldots, C$, with C being the total number of channels. $N$ is the sequence length, and $[x_1, \ldots, x_c, \ldots, x_C] \in R^{N \times C}$. As shown in Fig. 3.1c, a one-dimensional discrete wavelet transform is applied to each input signal channel $x_c$ up to a specific level L. Continuous decomposition is exclusively performed on low-frequency components derived from 1/2 down-sampling of intermediate variable sequences. At each level, $A_l$ represents the approximation coefficients of the signal, which are low-frequency components generated by a low-pass filter $\phi = [\phi_0, \ldots, \phi_m, \ldots, \phi_M]$, and $D_l$ represents the detail coefficients, which are high-frequency components from a high-pass filter $\varphi = [\varphi_0, \ldots, \varphi_m, \ldots, \varphi_M]$, with $M << N$. Subseries at each level are computed as:

$$A_{l+1}[k] = \sum_{m=0}^{M} x_l[k+m] \cdot \phi_{[m]}, \tag{3.1}$$

$$D_{l+1}[k] = \sum_{m=0}^{M} x_l[k+m] \cdot \varphi_{[m]}, \tag{3.2}$$

where $A_{l+1}$ and $D_{l+1}$ are sub-approximation and sub-detail coefficients, respectively. $x_0$ is the input series, and $x_l[k]$ is the k-th element of low-frequency components at level $l$, $l = 0, 1, \ldots, L$. Approximation coefficients represent smoothed averages of the input signal. We augment only detail coefficients to the original sequences, as approximation coefficients may constitute redundant data that impede model training. Consequently, the augmented segment for each channel is represented as $X_c = [x_c, d_c^1, \ldots, d_c^L]$.

#### 3.2.3.2 LSTM Branch

Wind turbine monitoring data are time series, representing discrete data with temporal dependencies rather than independent points. Extracting valuable information from original and transformed sequences is crucial. LSTM is used based on the understanding that temporal correlations within a time series are closely linked to frequency [26]. LSTM excels at processing and predicting significant events with long-term dependencies in time series. Additionally, a Dropout layer is integrated with the LSTM layer to enhance generalization. Based on the previous subsection, the input to the LSTM branch is defined as: $X_{in} = [X, D_1, \ldots, D_L] = [[x_0^t], \ldots, [x_l^t], \ldots, [x_L^t]]$, where $X_l \in R^{C\times \frac{N}{2^l}}$, and $t$ denotes the time stamps for each subsequence. The computation at each LSTM layer time step is described by Graves et al. [27], and reformulated here for consistency:

$$\begin{aligned}
g^i &= \sigma(W^i h^{t-1} + I^i x_l^t), \\
g^f &= \sigma(W^f h^{t-1} + I^f x_l^t), \\
g^o &= \sigma(W^o h^{t-1} + I^o x_l^t), \\
g^c &= tanh(W^c h^{t-1} + I^c x_l^t), \\
m^t &= g^f \otimes m^{t-1} + g^i \otimes g^c, \\
h^t &= tanh(g^o \otimes m^t), \\
y &= f_{dropout,p}(g^o),
\end{aligned} \tag{3.3}$$

where $g^i$, $g^f$, $g^o$, $g^c$ are activation vectors for input, forget, output, and cell state gates, respectively, and $\sigma$ is the logistic sigmoid function. Recurrent weight matrices are $W^i$, $W^f$, $W^o$, $W^c$, and $I^i$, $I^f$, $I^o$, $I^c$ are projection matrices. The hidden state vector is $h^t$, and $\otimes$ denotes element-wise multiplication. Dropout calculation is $f_{dropout,p}$, with $p$ being the probability of retaining a network unit.

#### 3.2.3.3 CNN Branch

CNN is utilized to capture compact data features from short-term contextual information in time series. The CNN branch comprises basic convolutional blocks, each with a convolutional layer, batch normalization layer, and ReLU layer. Convolution layers effectively merge temporal and channel information, ensuring thorough extraction of crucial features from raw data and detail coefficients for icing detection. Batch normalization (BN) accelerates training and improves generalization. ReLU layers reduce model overfitting by decreasing parameter interdependence. The CNN branch input is identical to the LSTM branch input. We denote $X_{in}$ from the LSTM branch as $X_0$ in the CNN branch. The basic convolution block is

$$X_i = ReLU(BN(f(W * X_{i-1} + b))), i = 0, 1, \ldots, I, \tag{3.4}$$

where $X_i$ is the input or output of each block, $I$ is the total number of blocks, $f$ is the activation function, and $*$ is convolution. $W$ and $b$ are learned convolutional layer parameters. Following convolutional blocks, a global average pooling layer reduces dimensionality and network parameters. Outputs from LSTM and CNN branches, and their combined outcomes, are merged via a concatenate layer.

## 3.2.4 *Icing Detection*

### 3.2.4.1 Icing Detection Classifier

Multilevel features learned by MCRNN, as shown in Fig. 3.1d, are input to a classifier for blade icing detection. The classifier computes the probability distribution indicating the likelihood of blade icing, enabling prediction of icing presence. In this work, a dense layer with a softmax function serves as the classifier. The dense layer applies a weight matrix and bias to map n real numbers from $(-\infty, +\infty)$ to K real numbers, and the softmax function maps these K real numbers to probabilities between 0 and 1, summing to 1, where K is the number of categories. The dense layer and softmax process is

$$\hat{y} = softmax(z) = softmax(W^T x + b), \tag{3.5}$$

$$softmax(z_j) = \frac{e^{z_j}}{\sum_{k=1}^{K} e^{z_k}}, j = 1, \ldots, K, \tag{3.6}$$

where $x$ is the dense layer input, $W$ and $b$ are weight and bias, and $\hat{y}$ is class probability. Training and validation sets are used to optimize models for versatility and generalization. Optimal models are then deployed for real-world icing detection.

### 3.2.4.2 Multi-step Accumulation Strategy

After offline training, optimal MCRNN models are ready for real-time blade icing detection. Real-time detection employs a sliding window of the same length as in offline training. The window moves across real-time data, and the trained model predicts icing probability for each window. However, variable turbine operating conditions and environmental factors can significantly alter monitoring signals, leading to detection errors. High false alarm rates are inefficient and complicate monitoring efforts. To improve accuracy and reduce false alarms, we propose a multi-step accumulation strategy. As shown in Fig. 3.3, for a predicted icing probability sequence $[p_1, p_2, \ldots, p_{n-1}, p_n]$, we compare each probability $p_m$ $(m < n)$ to a threshold $T$. We also examine results for $p_{m+1}$ to $p_{m+k-1}$. If $k$ consecutive probabilities exceed $T$, an icing warning is issued. The value of $k$ can be adjusted based on application needs and data fragment length.

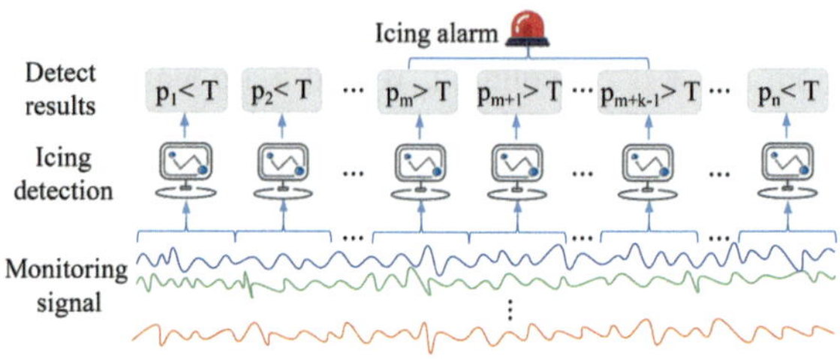


**Fig. 3.3** An illustration of the multi-step accumulation strategy. $p_n$ is the icing possibility and T is the threshold

### 3.2.5 Loss Function

Raw SCADA data exhibit significant imbalance, thus we investigate both data resampling and loss function-based methods for model optimization. For data resampling, raw data are resampled to create balanced datasets. Cross-entropy loss is used to optimize models trained on balanced data. For loss function-based optimization, imbalanced datasets are used directly, and the focal loss function is employed for optimization [30].

Cross-entropy loss (CL) is defined as:

$$CL = -y\log(\hat{y}) + (1-y)\log(1-\hat{y}), \tag{3.7}$$

where $y$ is the data label, with 1 for positive (icing) and 0 for negative (normal) samples. $\hat{y}$ is the predicted probability of a positive sample.

Focal loss (FL) [31] modulates cross-entropy loss to reduce the loss for well-classified samples, focusing on difficult samples. FL equation is

$$\begin{aligned} p_t &= \begin{cases} \hat{y}, & \text{if } y = 1, \\ 1-\hat{y}, & \text{otherwise.} \end{cases} \\ FL(p_t) &= -\alpha_t(1-p_t)^{\gamma}\log(p_t), \end{aligned} \tag{3.8}$$

where $\alpha_t$ balances positive/negative class importance, and $(1-p_t)^{\gamma}$ modulates simple/difficult samples, with $\gamma$ being the focusing parameter. Simple samples (from majority class) are down-weighted, focusing the loss function on difficult samples (from minority class). Focal loss mitigates issues from imbalanced data that cause models to perform poorly on under-represented classes (those with fewer data points).

## 3.3 Experiment

### *3.3.1 Experimental Setup*

This section details the experimental configuration, including data sources and evaluation metrics. We then describe data analysis and preprocessing, followed by an evaluation of the proposed MCRNN model through detailed experiments, and a discussion of the experimental outcomes.

**(1) SCADA Data**

Experimental data are from standard datasets provided by the Ministry of Industry and Information Technology of the People's Republic of China, including monitoring data from two wind turbines manufactured by Goldwind, Inc., located in Inner Mongolia, China. We recorded SCADA data over 695.59 h and 305.77 h for these turbines, respectively. The SCADA system gathers data from hundreds of sensors. Domain experts identified 26 variables related to blade icing, detailed in Table 3.1, with sensor locations shown in Fig. 3.4. Datasets from these turbines are named WT-1 and WT-2.

**Table 3.1** Wind turbine SCADA data parameters

| Index | Monitored variable | Index | Monitored variable |
|---|---|---|---|
| 1 | Wind velocity | 14 | Pitch motor 1 temperature |
| 2 | Generator rotational speed | 15 | Pitch motor 2 temperature |
| 3 | Active Electrical power output | 16 | Pitch motor 3 temperature |
| 4 | Wind Direction (Compass Heading) | 17 | Horizontal Acceleration (Turbine Structure) |
| 5 | 25-Second Averaged wind direction angle | 18 | Vertical Acceleration (Turbine Structure) |
| 6 | Yaw system position | 19 | Ambient temperature |
| 7 | Yawing rate | 20 | Nacelle internal temperature |
| 8 | Pitch angle - Blade 1 | 21 | Pitch 1 switching Element temperature |
| 9 | Pitch angle - Blade 2 | 22 | Pitch 2 switching Element temperature |
| 10 | Pitch angle - Blade 3 | 23 | Pitch 3 switching Element temperature |
| 11 | Pitch actuator 1 speed | 24 | Pitch 1 charger DC power Consumption |
| 12 | Pitch actuator 2 speed | 25 | Pitch 2 charger DC power consumption |
| 13 | Pitch actuator 3 speed | 26 | Pitch 3 charger DC power consumption |

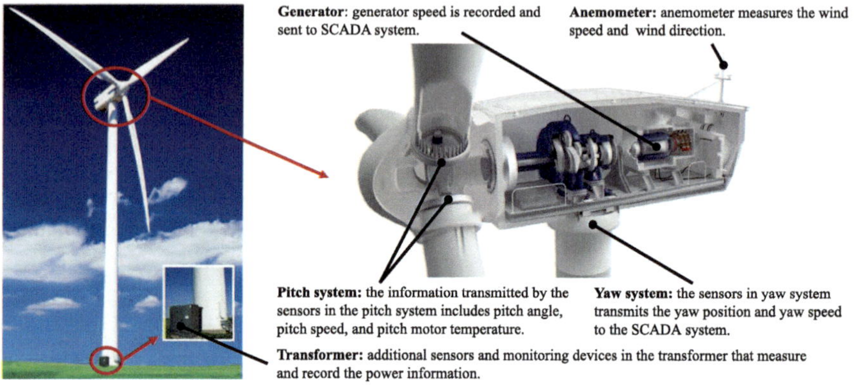


**Fig. 3.4** Sketch of sensor location in the components of a wind turbine (modified figures from [28, 29]). The specifications of the sensors are presented in Table 3.1

**(2) Evaluation Metrics**

Testing phases consistently use imbalanced datasets to evaluate model effectiveness. For imbalanced data, accuracy is unsuitable due to bias towards the majority class (normal operation). Therefore, **precision**, **recall**, and **F1-score** are used:

$$Precision = \frac{TP}{TP + FP}, \tag{3.9}$$

$$Recall = \frac{TP}{TP + FN}, \tag{3.10}$$

$$F1\text{-}score = \frac{2 \times Precision \times Recall}{Precision + Recall}, \tag{3.11}$$

where $TP$, $FP$, $FN$, and $TN$ are true positive, false positive, false negative, and true negative counts. Precision focuses on minimizing false positives, while recall emphasizes minimizing false negatives. F1-score balances precision and recall, serving as the primary evaluation metric.

**(3) Parameter Settings**

In MCRNN, MDWD level is set to $\{3\}$ (justified in Sect. 3.3-E). The LSTM branch has one LSTM layer with 16 hidden units and a dropout rate $p = 0.5$. The CNN branch has three blocks with filter sizes $\{64, 128, 64\}$, and 1D kernels of sizes $\{8, 5, 3\}$ without stride. Multi-step accumulation strategy steps $m = 3$ for real-time detection. Focal loss parameters $\alpha_t = 0.25$ and $\gamma = 3$. Settings are summarized in Table 3.2.

**Table 3.2** Parameters settings

| Setting description | Setting or value |
|---|---|
| MDWD level | 3 |
| Number of LSTM layer in LSTM branch | 1 |
| Number of hidden units in LSTM layer | 16 |
| $p$ in Dropout layer | 0.5 |
| Number of basic convolution block in CNN branch | 3 |
| The filter sizes of the convolutional layers | 64, 128, 64 |
| The kernel sizes in convolutional layers | 8, 5, 3 |
| $\alpha_t$ in FL function | 0.25 |
| $\gamma$ in FL function | 3 |

## 3.3.2 Data Analysis and Processing

**(1) Data Analysis**

The wind turbine's core is the electrical machinery generating power from wind. Icing alters blade aerodynamics, reducing power output at a given wind speed. Figure 3.5 shows that for a given wind speed, power significantly decreases after ice accumulation compared to normal operation. Note that raw data is transformed for privacy. Similar to power, generator speed also decreases with icing. As wind speed increases, power plateaus; data beyond this point (wind speed $>2$, generator speed $<1.5$) are redundant and discarded as they do not contribute to icing detection and may degrade feature representation.

Mechanically, wind turbine blades are balanced for weight and torque to prevent fatigue. Icing disrupts this balance, potentially causing imbalance. However, examining variables related to the three blades shows consistency even with icing. To

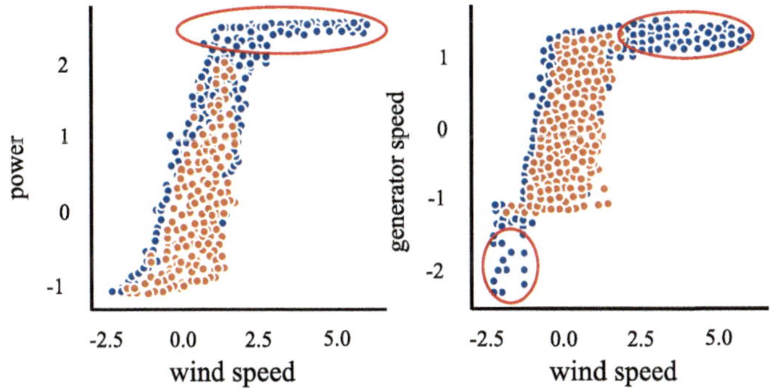


**Fig. 3.5** Data distribution of wind speed versus power, and wind speed versus generator speed. (Blue dots represent normal data, and orange dots represent icing data). The visualized data are normalized due to the confidentiality of the original data, resulting in the loss of data units

mitigate overfitting, we averaged variables with the same attributes across the three blades (Table 3.1) for experiments.

**(2) Data Processing**
Raw signals are segmented into fixed-length fragments for model input. Window sizes for WT-1 and WT-2 are {32} and {64}, respectively. Each fragment is labeled as icing or normal. To address data imbalance, we augmented positive samples (icing data) by generating overlapping abnormal fragments and non-overlapping normal fragments (Fig. 3.2a). This resulted in two datasets: balanced (resampled) and imbalanced (original).

### *3.3.3 Baseline Comparison*

We compared our model with four baselines on both balanced and imbalanced datasets:

- **LSTM**: Long Short-Term Memory network, designed for long-term dependencies in time series.
- **MLSTM-FCN**: Multivariate LSTM Fully Convolutional Network, efficient for real-time systems [32].
- **FCNN**: Fully Convolutional Neural Network for time-series classification [33].
- **WaveletFCNN**: Deep learning model for wind turbine blade icing detection using wavelet transformation [23].
- **SSENET**: Strong baseline for multivariate time-series classification using dense connections. Attention mechanism removed for fair comparison of model structure [18].

MCRNN$^C$ uses cross-entropy loss, and MCRNN$^F$ uses focal loss, applied to both balanced and imbalanced datasets.

Results on balanced datasets are in Fig. 3.6. MCRNN$^C$ improves F1-score on WT-1 by 34.8% and on WT-2 by 42.9% over the best baseline (SSENET). MCRNN$^F$ improves F1-score on WT-1 by 23.9% and WT-2 by 30.6% over SSENET. Cross-entropy loss performs better than focal loss on balanced data. Baselines show high recall but low precision, indicating many false positives, unsuitable for real-world icing detection due to high false alarm rates.

Results on imbalanced datasets are in Fig. 3.7. Proposed models achieve highest F1-scores. MCRNN$^C$ improves by 1.8% and 5.7% over SSENET on WT-1 and WT-2, respectively. MCRNN$^F$ improves by 14.3% and 34%, respectively. MCRNN$^F$ is the best model for imbalanced data, with F1-scores 12.3 and 26.8% higher than MCRNN$^C$, demonstrating focal loss's effectiveness in mitigating class imbalance.

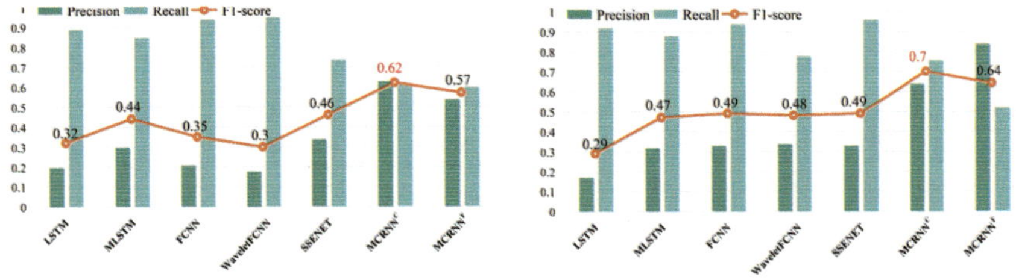


(a) WT-1 (b) WT-2

**Fig. 3.6** Baseline comparison on balanced datasets of **a** WT-1, and **b** WT-2

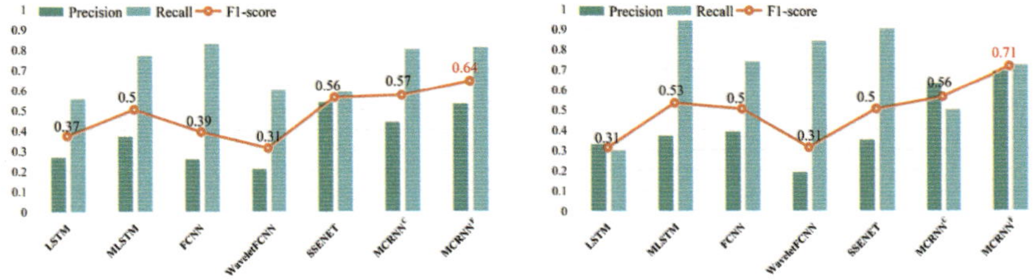


(a) WT-1 (b) WT-2

**Fig. 3.7** Baseline comparison on imbalanced datasets of **a** WT-1, and **b** WT-2

## *3.3.4 Comparison to Different Model Variants*

We compared three MCRNN variants to assess component contributions:

- **MCRNN-MR**: MCRNN with MDWD and LSTM, no CNN.
- **MCRNN-MC**: MCRNN with MDWD and CNN, no LSTM.
- **MCRNN-CR**: MCRNN without MDWD.

Table 3.3 shows results on balanced datasets. Removing any module reduces performance. For example, MCRNN$^C$ and MCRNN$^F$ outperform MCRNN-CR$^C$ and MCRNN-CR$^F$ significantly on both datasets. Table 3.4 shows similar trends on imbalanced datasets. MCRNN$^C$'s F1-scores are 36.8 and 60.6% higher on average than variants on WT-1 and WT-2, respectively; MCRNN$^F$'s are 25.2 and 37.2% higher. LSTM and CNN modules are both crucial for performance, highlighting temporal dependency and CNN's feature extraction capabilities.

Table 3.4 shows variant comparisons on imbalanced data. Removing any component degrades performance. Focal loss consistently improves F1-score over cross-entropy loss for all variants on imbalanced data, confirming its benefit for icing detection under class imbalance.

**Table 3.3** Comparison of different model variants on balanced datasets

| | | MCRNN-MR$^C$ | MCRNN-MC$^C$ | MCRNN-CR$^C$ | MCRNN$^C$ | MCRNN-MR$^F$ | MCRNN-MC$^F$ | MCRNN-CR$^F$ | MCRNN$^F$ |
|---|---|---|---|---|---|---|---|---|---|
| WT-1 | *Precision* | 29 | 33 | 33 | 63 | 29 | 31 | 42 | 54 |
| | *Recall* | 91 | 79 | 75 | 61 | 84 | 80 | 58 | 60 |
| | *F1-score* | 44 | 46 | 46 | **62** | 43 | 45 | 49 | **57** |
| WT-2 | *Precision* | 27 | 31 | 30 | 64 | 30 | 44 | 37 | 84 |
| | *Recall* | 88 | 86 | 90 | 76 | 94 | 52 | 64 | 52 |
| | *F1-score* | 41 | 45 | 45 | **70** | 45 | 48 | 47 | **64** |

**Table 3.4** Comparison of different model variants on imbalanced datasets

| | | MCRNN-MR$^C$ | MCRNN-MC$^C$ | MCRNN-CR$^C$ | MCRNN$^C$ | MCRNN-MR$^F$ | MCRNN-MC$^F$ | MCRNN-CR$^F$ | MCRNN$^F$ |
|---|---|---|---|---|---|---|---|---|---|
| WT-1 | *Precision* | 33 | 30 | 20 | 44 | 29 | 29 | 24 | 53 |
| | *Recall* | 30 | 66 | 90 | 80 | 84 | 76 | 85 | 81 |
| | *F1-score* | 31 | 41 | 32 | **57** | 43 | 43 | 37 | **64** |
| WT-2 | *Precision* | 29 | 43 | 18 | 63 | 44 | 56 | 43 | 69 |
| | *Recall* | 76 | 40 | 58 | 50 | 52 | 60 | 40 | 72 |
| | *F1-score* | 42 | 42 | 28 | **56** | 48 | 57 | 42 | **71** |

### 3.3.5 Sensitivity Analysis

Two sensitivity analyses were conducted: (1) MDWD level sensitivity; (2) Window size sensitivity.

#### 3.3.5.1 Impact of Different MDWD Level

MDWD level variations can affect analysis results. We analyzed MDWD levels 2, 3, 4, and 5 to find the optimal level for icing detection. Decomposition is applied to down-sampled low-frequency components, so excessively deep decomposition can reduce frequency resolution due to short sequence lengths at deeper levels.

Experiments used balanced datasets with cross-entropy loss and imbalanced datasets with focal loss. Table 3.5 shows MDWD level sensitivity results. Level 3 consistently yields the highest F1-scores across models and datasets, indicating it captures better time and frequency domain information for feature extraction.

**Table 3.5** MDWD level sensitive analysis

| | | MDWD Level | | | |
|---|---|---|---|---|---|
| | | 2 | 3 | 4 | 5 |
| WT-1 Balanced data & CL function | *Precision* | 45 | 63 | 45 | 41 |
| | *Recall* | 72 | 61 | 80 | 71 |
| | *F1-score* | 55 | **62** | 59 | 52 |
| WT-1 Imbalanced data & FL function | *Precision* | 62 | 53 | 58 | 65 |
| | *Recall* | 64 | 81 | 64 | 52 |
| | *F1-score* | 63 | **64** | 61 | 59 |
| | | | | | |
| WT-2 Balanced data & CL function | *Precision* | 49 | 64 | 55 | 49 |
| | *Recall* | 86 | 76 | 82 | 78 |
| | *F1-score* | 62 | **70** | 66 | 60 |
| WT-2 Imbalanced data & FL function | *Precision* | 56 | 69 | 66 | 43 |
| | *Recall* | 66 | 72 | 42 | 84 |
| | *F1-score* | 61 | **71** | 51 | 57 |

#### 3.3.5.2 Impact of Different Window Size

Window size sensitivity was analyzed using window sizes 32, 64, 128, and 256 (approximately 3.5, 7, 15, and 30 minutes of data) on both balanced and imbalanced datasets.

Figure 3.8 shows balanced dataset results. Window size 32 is optimal for WT-1, and 64 for WT-2. Shorter window sizes (32 or 64) allow earlier icing detection. Figure 3.9 shows imbalanced dataset results, where sizes 32 and 64 are also superior, consistent with balanced dataset findings.

### *3.3.6 Real-Time Icing Detection*

To validate practical application, real-time icing detection was performed on part of the WT-1 test dataset. Data segments of fixed length (same as training) were collected for real-time detection. A sliding window (length equal to training window size) moved with a stride of 8 (approximately one minute), enabling minute-by-minute detection. Optimal models from training Table 3.6 were used. Icing/normal probability threshold was set to 0.5. Thresholds in real-world scenarios should be

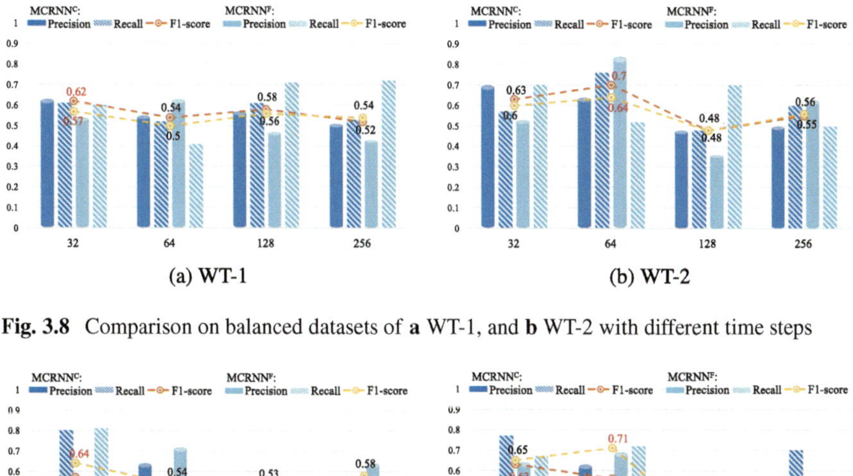


(a) WT-1 (b) WT-2

**Fig. 3.8** Comparison on balanced datasets of **a** WT-1, and **b** WT-2 with different time steps

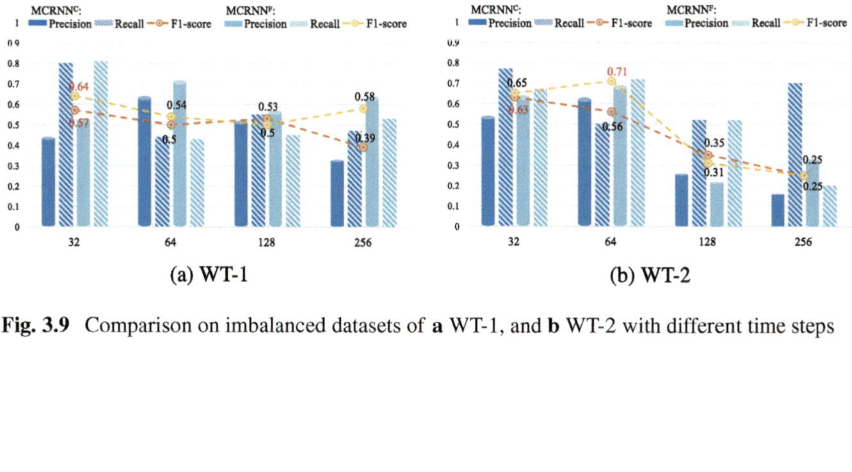


(a) WT-1 (b) WT-2

**Fig. 3.9** Comparison on imbalanced datasets of **a** WT-1, and **b** WT-2 with different time steps

**Table 3.6** Information of optimal models

| Model | Training data | Model | F1-score | Precision | Recall |
|---|---|---|---|---|---|
| I | WT-1 | $MCRNN^C$ | 62 | 63 | 61 |
| II | WT-1 | $MCRNN^F$ | 64 | 53 | 81 |

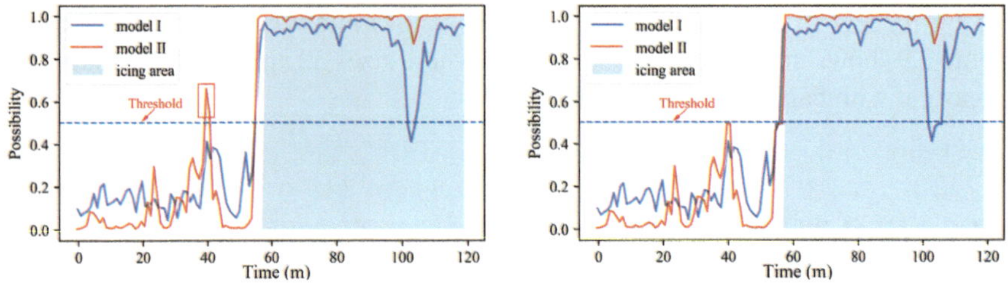


(a) WT-1 (b) WT-2

**Fig. 3.10** Real-time detection on WT-1 data with **a** directly detection and **b** detection with multi-step accumulation strategy. (I, II) indicate the optimal models (listed in Table 3.6) employed for real-time icing detection, and the information of the models is listed in Table 3.6

refined based on observed probability-ice amount relationships. Multi-step accumulation strategy ($k = 3$) was used to improve robustness and reduce false alarms.

Figure 3.10 shows results with direct detection (a) and multi-step accumulation (b). Both models detected all icing areas. Direct detection Fig. 3.10a shows false alarms (red box) in non-icing periods. Multi-step accumulation Fig. 3.10b significantly reduces false alarms, effectively lowering false alarm rates in non-icing periods.

### *3.3.7 Discussion*

Experiments on balanced and imbalanced datasets show MCRNN outperforms state-of-the-art approaches in F1-score Figs. 3.6 and 3.7, demonstrating its effectiveness for multisensor time-series icing detection. For data imbalance, $MCRNN^C$ on balanced data achieved highest F1-score, 16.9% higher on average than on no-sampling datasets, suggesting resampling improves multivariate time-series correlation capture. $MCRNN^F$ performed best on imbalanced datasets, outperforming data resampling by 1.4–3.2% in F1-score. Experiments validate both imbalance solutions, particularly focal loss's effectiveness for imbalanced multivariate time series. MDWD level 3 is optimal for MCRNN. Shorter time steps improve detection, aligning with real-world needs for early detection. Real-time detection shows multi-step accumulation reduces false alarms, improving robustness and reducing monitoring personnel workload.

## 3.4 Conclusion

This paper introduces MCRNN to address challenges in wind turbine blade icing detection, capturing multi-channel correlations and multi-scale dynamic patterns in time-series data. MCRNN has two core modules: (1) Data transformation using multilevel discrete wavelet decomposition to extract time and frequency domain information crucial for feature learning. (2) Effective use of LSTM for temporal dependency analysis and CNN for powerful feature extraction. MCRNN maximizes time-frequency information from raw and MDWD-transformed data. Data and algorithm-level solutions address data imbalance. Multi-step accumulation enhances robustness. Extensive experiments validate MCRNN's generalization, effectiveness of imbalance solutions, and necessity of multi-step accumulation. Results show MCRNN outperforms several methods for wind turbine blade icing detection.

## References


1. Ren Z, Verma AS, Li Y, Teuwen JJ, Jiang Z (2021) Offshore wind turbine operations and maintenance: a state-of-the-art review. Renew Sustain Energy Rev 144:110886
2. Top 10 wind turbine manufacturers in the world. https://www.bizvibe.com/blog/top-10-wind-turbine-manufacturers-world/. Accessed 03 Feb 2019
3. Wei K, Yang Y, Zuo H, Zhong D (2020) A review on ice detection technology and ice elimination technology for wind turbine. Wind Energy 23
4. Madi E, Pope K, Huang W, Iqbal T (2019) A review of integrating ice detection and mitigation for wind turbine blades. Renew Sustain Energy Rev 103:269–281
5. Wang P, Zhou W, Bao Y, Li H (2018) Ice monitoring of a full-scale wind turbine blade using ultrasonic guided waves under varying temperature conditions. Struct Control Health Monit 25(4):e2138.1–e2138.17
6. Carlsson V (2010) Measuring routines of ice accretion for wind turbine applications: the correlation of production losses and detection of ice. Master's thesis, Department of Physics, Faculty of Science and Technology, Umeå University, 2010
7. Bin XU, Guo X, Chen H (2017) Active icing monitoring for wind turbine blade models with pzt technology. Yadian Yu Shengguang/Piezoelectrics and Acoustooptics 39(1):72–76
8. Roberge P, Lemay J, Ruel J, Bégin-Drolet A (2018) A new atmospheric icing detector based on thermally heated cylindrical probes for wind turbine applications. Cold Regns Sci Technol 148:131–141
9. Hsu DK, Margetan FJ, Wormley SJ, Simpson JA (1992) Apparatus and method for detection of icing onset and ice thickness, 17 Mar 1992, uS Patent 5,095,754
10. Laakso T, Holttinen H, Ronsten G, Tallhaug L, Horbaty R, Baring-Gould I, Lacroix A, Peltola E, Tammelin B (2003) State-of-the-art of wind energy in cold climates. IEA Annex XIX 24:53
11. Corradini ML, Cristofaro A, Pettinari S 2016) "A model-based robust icing detection and estimation scheme for wind turbines. European control conference (ECC). IEEE, 1451–1456
12. Shi Q (2017) Model-based detection for ice on wind turbine blades, Master's thesis, Department of Marine Technology, NTNU
13. Hu L, Zhu X, Chen J, Shen X, Du Z (2018) Numerical simulation of rime ice on NREL phase vi blade. J Wind Eng Ind Aerodyn 178:57–68
14. Song P, Yao Z, Zhao C (2019) Section division and multi-model method for early detection of icing on wind turbine blades. In: 34rd Youth Academic annual conference of chinese association of automation (YAC). IEEE, pp 749–754

15. Skrimpas GA, Kleani K, Mijatovic N, Sweeney CW, Jensen BB, Holboell J (2016) Detection of icing on wind turbine blades by means of vibration and power curve analysis. Wind Energy 19(10):1819–1832
16. Guangfei Z, Wen T, Da Z (2018) Ice detection for wind turbine blades based on PSO-SVM method. J Phys: Conf Ser 1087:022036
17. Yi H, Jiang Q, Yan X, Wang B (2020) Imbalanced classification based on minority clustering smote with wind turbine fault detection application. IEEE Trans Ind Inf
18. Cheng X, Li G, Ellefsen AL, Chen S, Hildre HP, Zhang H (2020) A novel densely connected convolutional neural network for sea-state estimation using ship motion data. IEEE Trans Instrum Meas 69(9):5984–5993
19. Ellefsen AL, Æsøy V, Ushakov S, Zhang H (2019) A comprehensive survey of prognostics and health management based on deep learning for autonomous ships. IEEE Trans Reliab 68(2):720–740
20. Chen L, Xu G, Zhang Q, Zhang X (2019) Learning deep representation of imbalanced SCADA data for fault detection of wind turbines. Measurement 139:370–379
21. Liu Y, Cheng H, Kong X, Wang Q, Cui H (2019) Intelligent wind turbine blade icing detection using supervisory control and data acquisition data and ensemble deep learning. Energy Sci. Eng. 7(6):2633–2645
22. Cheng X, Shi F, Zhao M, Li G, Zhang H, Chen S (2021) Temporal attention convolutional neural network for estimation of icing probability on wind turbine blades. IEEE Trans Ind Electron 1–1
23. Yuan B, Wang C, Jiang F, Long M, Yu PS, Liu Y (2019) Waveletfcnn: a deep time series classification model for wind turbine blade icing detection. arXiv: abs/1902.05625
24. Cui Y, Jia M, Lin T-Y, Song Y, Belongie S (2019) Class-balanced loss based on effective number of samples. In: Proceedings of the IEEE conference on computer vision and pattern recognition, pp 9268–9277
25. Wang X, Tang G, Wang T, Zhang X, Peng B, Dou L, He Y (2020) Lkurtogram guided adaptive empirical wavelet transform and purified instantaneous energy operation for fault diagnosis of wind turbine bearing. IEEE Trans Instrum Meas 70:1–19
26. Wang J, Wang Z, Li J, Wu J (2018) Multilevel wavelet decomposition network for interpretable time series analysis. In: Proceedings of the 24th ACM SIGKDD international conference on knowledge discovery and data mining, pp 2437–2446
27. Kawakami K (2008) Supervised sequence labelling with recurrent neural networks, PhD dissertation, Department of Electrical and Electronics Engineering, Technical University Munich, Munich, Germany
28. Wind turbine step-up transformers. https://studylib.net/doc/18480053/wind-turbine-step-up-transformers. Accessed 09 Mar 2020
29. Predictive maintenance of wind turbines and generators. https://www.maintenanceandengineering.com/2019/08/12/predictive-maintenance-of-wind-turbines-and-generators/. Accessed 09 Mar 2020
30. Han P, Li G, Skulstad R, Skjong S, Zhang H (2021) A deep learning approach to detect and isolate thruster failures for dynamically positioned vessels using motion data. IEEE Trans Instrum Meas 70:1–11
31. Lin T-Y, Goyal P, Girshick R, He K, Dollár P (2017) Focal loss for dense object detection. In: Proceedings of the IEEE international conference on computer vision, pp 2980–2988
32. Karim F, Majumdar S, Darabi H, Harford S (2018) Multivariate LSTM-FCNS for time series classification. Neural Netw 237–245
33. Wang Z, Yan W, Oates T (2017) "Time series classification from scratch with deep neural networks: a strong baseline. Int Joint Conf Neural Netw (IJCNN) 2017:1578–1585

# Chapter 4
# Class-Imbalanced Semi-supervised Temporal Neural Network for Blade Icing Detection

## 4.1 Introduction

Wind energy constitutes a pivotal element in global strategies aimed at mitigating climate change and fostering the transition towards sustainable energy paradigms. Recent years have witnessed an exponential surge in the installed capacity of wind turbines. To optimize the utilization of wind resources, wind farms are increasingly located in regions characterized by elevated altitudes and latitudes [1]. However, the deployment of wind turbines in such environments frequently precipitates blade icing, a phenomenon with deleterious repercussions on aerodynamic efficiency, site accessibility due to ice ejection hazards, aggregate annual energy yield, and the operational lifespan of turbines [2]. The accretion of ice on turbine blades not only compromises power generation, potentially incurring losses of up to 30% in annual energy production under severe icing conditions, but also engenders safety hazards in the vicinity of wind power installations [3]. These operational and safety imperatives have stimulated extensive research into methodologies for the prompt identification and subsequent mitigation of blade ice accumulation [4]. Consequently, the precise and timely detection of blade icing on wind turbines is of paramount importance for ensuring the efficacious and secure operation of wind farms.

Presently, human observation remains a prevalent approach for blade icing identification within wind farm settings. This subjective methodology entails the assessment of icing conditions through the interpretation of wind turbine power curves, in conjunction with environmental sensor readings and rotor speeds. While human observation offers the merit of consistency and independence from sole reliance on external sensor systems, its intrinsic subjectivity presents a notable limitation. Discrepancies in observer interpretation can lead to inconsistent evaluations and may result in the failure to detect critical levels of blade icing. To address the inherent shortcomings of human observation, a spectrum of anti-blade-icing and de-blade-icing technologies have been explored, broadly categorized into passive, active, and hybrid modalities. Passive strategies aim to preempt ice formation via the applica-

 
X. Cheng et al., *Computational Methods for Blade Icing Detection of Wind Turbines*, Engineering Applications of Computational Methods 24,
https://doi.org/10.1007/978-981-96-6763-5_4

tion of specialized materials [5], encompassing liquid-infused surfaces and porous superhydrophobic polyvinylidene fluoride coatings [6, 7]. These passive techniques offer the benefits of diminished operational expenditures and autonomous ice prevention capabilities [8]. Conversely, active methodologies employ controllable thermal or mechanical mechanisms to eliminate or prevent icing, including electro-thermal and ultrasonic de-icing apparatuses widely implemented in wind farms [9, 10]. While affording controllability, active methods necessitate supplementary mechanical maintenance and component replacement cycles. Hybrid methodologies endeavor to synthesize the advantages of both active and passive approaches, garnering increasing scholarly and industrial interest [2, 11]. Nevertheless, these conventional techniques frequently entail substantial financial investments and demand significant manual labor. Critically, their diagnostic accuracy is constrained as they estimate icing conditions without comprehensively accounting for the turbine's intrinsic operational variabilities. Furthermore, mechanical interventions associated with component replacement or novel installations can potentially compromise the structural integrity of the turbine [12].

In parallel with these physical and mechanical interventions, substantial research endeavors have concentrated on blade icing detection grounded in the fundamental physical processes governing icing formation [13, 14]. Icing models are broadly bifurcated into empirical statistical models (ESM) and icing growth models (IGM). ESMs are contingent upon longitudinal environmental monitoring and historical data analytics to forecast ice accretion frequency and volume within specific geographical areas. However, ESMs often exhibit suboptimal accuracy. IGMs, in contrast, conceptualize blade icing as a complex physical phenomenon encompassing aerodynamics, multiphase gas-liquid hydrodynamics, and heat transfer dynamics. While IGMs furnish valuable insights into the icing process, they necessitate considerable domain-specific expertise [15] and are heavily reliant on specific assumptions and auxiliary experimental infrastructure such as wind tunnels [16].

The pervasive integration of sensor technologies within wind turbines has facilitated the accumulation of extensive datasets encompassing both normal operational states and anomalous conditions. This data availability has catalyzed burgeoning interest in data-driven methodologies for blade icing detection, which endeavor to extract latent information from historical operational data [17]. Data-driven approaches present compelling advantages over traditional methodologies by obviating the prerequisite for prior domain knowledge, specialized materials, or mechanical tools. Data-driven strategies can be further sub-classified into machine learning and deep learning-based techniques. Machine learning paradigms typically involve the extraction of salient features that characterize icing conditions and the subsequent development of classification models, frequently employing shallow architectures such as logistic regression, support vector machines (SVM), and random forests [18, 19]. For example, Jimenez et al. introduced a machine learning model specifically designed for detecting ice presence and quantifying its thickness [18]. While efficacious, feature extraction is a computationally intensive process, demands domain-specific proficiency, and is often decoupled from the icing detection pipeline, potentially impacting holistic system performance. In recent years, deep learning has emerged as a potent analytical tool, demonstrating notable success across diverse domains, including wind turbine blade icing identification [12, 20]. Deep-learning-based methodologies, such as the ensemble deep learning framework predicated on multilevel SCADA data features articulated by Liu et al. [20], and the wavelet-CNN architecture proposed by Yuan et al. [12], present a promising and effective resolution by autonomously learning high-level data representations and discerning icing conditions via hierarchical network structures [21]. These techniques capitalize on the superior automated feature learning proficiencies inherent to deep learning, which are crucial for robust blade icing identification.

Notwithstanding these advancements, substantial challenges persist in the effective application of deep learning to blade icing detection. Principally, deep learning models conventionally require voluminous quantities of annotated data for robust training. The acquisition of labeled data for wind turbine icing constitutes a major impediment, as data annotation is both financially and temporally prohibitive. A significant proportion of collected wind farm data remains unlabeled. This data scarcity issue is pervasive across numerous applications, including anomaly detection, wherein the objective is to pinpoint infrequent events that deviate markedly from the data majority [22]. Anomaly detection methodologies are categorized based on labeling prerequisites: fully supervised, semi-supervised/weakly supervised, and unsupervised. Fully supervised approaches, while effective, necessitate extensive labeled data resources, whereas unsupervised methods, such as Isolation Forest [23], Local Outlier Factor [24], and REPEN [25], obviate the need for labels but may compromise diagnostic accuracy. Semi-supervised/weakly supervised techniques, including DevNet [26], Deep SAD [27] and PReNet [26], offer a pragmatic middle ground when labeled data are partially available, aligning with the inherent characteristics of wind turbine operational data. Secondarily, a pronounced class imbalance exists between ice-free (normal) and icing states, as turbines operate predominantly in non-icing regimes. This imbalance, often of unknown magnitude within unlabeled data, further complicates the analytical challenge. Thirdly, wind turbines function under heterogeneous environmental and climatic conditions, resulting in highly non-linear and non-stationary sensor data streams, rendering feature extraction for icing detection models particularly arduous. Time series classification (TSC), encompassing distance-based methodologies [28], feature-based approaches such as bag-of-features [29] and BOSS [30], and deep learning-based methods including multi-channel deep learning models [31] and LSTM-FCN architectures [17, 32, 33], provides a relevant framework for the analysis of wind turbine sensor data. However, the direct application of supervised TSC to blade icing detection is constrained by the scarcity of labeled data.

To address these multifaceted challenges, particularly the constraints imposed by labeled data scarcity and class imbalance, and drawing inspiration from semi-supervised learning (SSL) and class-imbalanced learning methodologies, this paper introduces a novel Deep Class-Imbalanced Semi-Supervised (DCISS) model specifically designed for wind turbine blade icing detection. Existing SSL techniques for

TSC frequently concentrate on univariate learning scenarios [34–36], with methods such as Dynamic Time Warping (DTW) [36] and frameworks like SUCCESS [37] and SSL architectures grounded in CNN and prototype learning [38] being developed. However, many SSL approaches operate under assumptions that either unlabeled data classes are congruent with labeled data classes [39] or that classes are intrinsically balanced [40, 41], neither of which is typically valid in blade icing detection contexts. To circumvent these limitations, the DCISS model leverages a semi-supervised learning (SSL) framework to effectively harness unlabeled data. To mitigate class imbalance effects, a prototypical network-based mechanism is implemented for class re-balancing. Prototypes for each class are constructed from the available labeled data, and class imbalance is addressed through their utilization [42]. Unlabeled data are then incorporated to refine these prototypes by evaluating the similarity between unlabeled data prototypes and labeled data prototypes, guided by the model training process. Furthermore, a channel attention module is integrated within the convolutional neural network (CNN) architecture to enhance feature extraction from complex training datasets. In contrast to pseudo-labeling or self-training methodologies in class-imbalanced SSL [43], our prototypical network offers a unified framework for concurrently addressing both class imbalance and semi-supervised learning imperatives.

In summation, this paper makes the following principal contributions:

- We propose DCISS, an innovative end-to-end deep learning framework that seamlessly integrates SSL and class-imbalanced learning methodologies. This framework effectively addresses class imbalance and fully exploits both labeled and unlabeled data through a prototypical learning paradigm. A channel calibration attention module is incorporated into the CNN to bolster feature extraction capabilities. This SSL methodology does not require a priori labeling knowledge, thereby offering broad applicability across diverse machine learning domains.
- We conduct a comprehensive evaluation of DCISS employing real-world SCADA data, contrasting its performance against state-of-the-art SSL and imbalanced learning techniques for blade icing detection. Empirical results unequivocally demonstrate the superior efficacy of DCISS. Online testing further corroborates the effectiveness, generalizability, and practicality of DCISS for real-time blade icing detection applications.

The subsequent sections of this paper are structured as follows: Sect. 4.2 provides a detailed exposition of the proposed DCISS model. Section 4.3 delineates the experimental evaluation protocol. Section 4.4 engages in a critical discussion of the results and Sect. 4.5 concludes the paper, summarizing key findings and outlining future research directions.

## 4.2 Methodology

This section commences with an overview of the proposed class-imbalanced SSL framework designed for blade icing detection in wind turbines. Subsequently, we elaborate on the granular details of each constituent component of the framework, encompassing data preprocessing, class-imbalanced SSL mechanisms, the feature encoder architecture, and prototype learning algorithms.

### 4.2.1 Framework Overview

As depicted in Fig. 4.1, the overarching framework is structured into three sequential phases: data preprocessing, DCISS model application, and blade icing detection assessment. Data preprocessing is designed to enhance the intrinsic quality of the data to facilitate robust modeling. The refined data are then input into the DCISS framework for class-imbalanced SSL. Finally, the trained DCISS model is deployed to evaluate blade icing conditions on wind turbines. Based on the resultant diagnostic outputs, proactive maintenance protocols or early icing prediction strategies can be implemented accordingly.

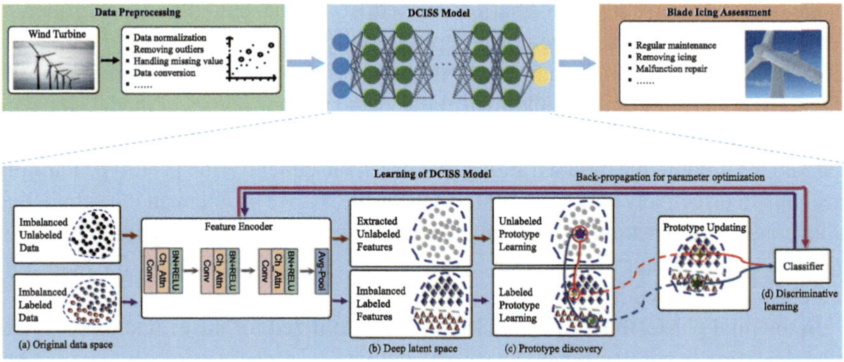


**Fig. 4.1** Schematic representation of the proposed DCISS model for blade icing detection in wind turbines. The forward propagation during DCISS training is indicated by the left-to-right arrow. **a** Original data space illustrating class-imbalanced labeled and unlabeled data (minority positive and majority negative classes); **b** Data distribution within a projected deep latent space achieved by the feature encoder; **c** Prototypical discovery via the prototypical network component; and **d** Discriminatory classifier learning by evaluating Euclidean distances between features and derived prototypes. Parameter optimization is executed through an inverse path (right-to-left) employing back-propagation of discriminative loss gradients to iteratively refine parameters within both the feature encoder and the discriminative classifier

To mitigate the potentially confounding influence of outliers and noise artifacts, a preliminary data preprocessing stage is indispensable for refining the raw time-series data. This preprocessing pipeline incorporates cleansing operations, correlation analysis, dataset partitioning, and normalization procedures. Correlation analysis is employed to scrutinize the interrelationships among the constituent time-series variables, with the objective of diminishing redundant information content. In this study, Pearson correlation analysis is specifically utilized for this purpose. Ultimately, the preprocessed time-series data are segmented into fixed-size temporal windows to facilitate the training of the deep learning model.

The primary objective of DCISS is to maximally leverage the available data resources, encompassing both labeled and unlabeled instances, for model training, while simultaneously addressing the inherent class imbalance challenge. Given that wind turbines typically operate under normal conditions for the majority of their operational lifespan, the acquired data are intrinsically class-imbalanced, i.e., the volume of samples representing the normal operational state significantly outweighs that of the icing state. It is pertinent to note that class imbalance is manifested in both the labeled and unlabeled data partitions. Critically, there is no a priori guarantee that the degree of imbalance will be consistent across these two data subsets. In this work, we introduce a class-imbalanced SSL methodology predicated on class prototypes to effectively address the imbalance issue within an SSL context. Within this approach, prototypes are initially derived for each class using the labeled data. Subsequently, the class imbalance is re-balanced through prototype-guided mechanisms [42]. Concurrently, information derived from unlabeled data is integrated to iteratively refine these balanced prototypes initially derived from labeled data. This refinement is achieved by quantifying the similarity between the prototype of the unlabeled data (representing unknown classes) and the prototypes of the labeled data (representing known classes). This similarity metric is dynamically computed and updated throughout the model training process. Consequently, the prototype learning mechanism effectively addresses both the class imbalance problem and the challenge of leveraging information from unlabeled data. We amalgamate these functionalities within a unified framework to provide enhanced, synergistic support for both SSL and class-imbalanced learning objectives.

In the final phase, the trained DCISS model is utilized for the evaluation of blade icing detection in wind turbines. The model outputs encompass a probabilistic assessment of icing conditions. An operator can then specify a threshold value (typically 0.5 is employed as a conventional default) to dichotomize the output and determine whether blade icing is present or absent. Thus, online evaluation capabilities empower operators to ascertain blade conditions in near real-time and undertake appropriate interventions, such as activating anti-icing/de-icing systems or initiating turbine shutdown protocols.

### 4.2.2 Deep Class-Imbalanced Semi-supervised (DCISS) Model

The SSL paradigm within our framework can be formally defined as follows: Given a training dataset comprising $N_l$ labeled samples, denoted as $S_l = \{x_1, x_2, \cdots, x_{N_l}\}$, their corresponding labels $Y_l = \{y_1, y_2, \cdots, y_{N_l}\}$, and $N_u$ unlabeled samples, $S_u = \{x_{N_l+1}, x_{N_l+2}, \cdots, x_{N_l+N_u}\}$, where typically $N_u >> N_l$, $x_i \in \mathbb{R}^D$ represents an individual sample with $D$ dimensions, $y_i \in \{1, 2 \cdots, n_c\}$ is the set of class labels for the labeled data, and $n_c$ denotes the number of distinct classes. The objective of SSL is to train a model $f([S_l; S_u]; \theta) \rightarrow Y_l$, parameterized by $\theta$ and learned from the composite training dataset. This learning task is mathematically framed as minimizing the generalization risk $R(f) = E_{([S_l;S_u],Y_l)}[L(f([S_l; S_u]; \theta), Y_l)]$, where $L$ represents the chosen loss function.

A prevalent assumption in many traditional SSL algorithms is that labeled and unlabeled samples exhibit approximately similar class distributions. That is, the number of samples across different classes is presumed to be roughly equivalent. However, when this assumption is violated, the decision boundaries learned by conventional SSL algorithms can become suboptimal, thereby degrading overall model performance. Exacerbating this issue is the potential for disparate degrees of class imbalance between labeled and unlabeled data subsets. Two primary strategic avenues exist to address the class imbalance problem: data distribution re-balancing and class-balanced loss methodologies. Data distribution re-balancing techniques aim to rectify class imbalance by oversampling minority class instances or undersampling majority class instances [44, 45]. However, a salient drawback of oversampling/undersampling is its inherent computational complexity, and performance can be markedly sensitive to dataset size variations, leading to protracted computation times. In contrast, class-balanced loss methods seek to re-weight loss contributions to align with a specified class distribution, thereby enhancing the generalization capacity of the model, particularly for minority classes. These methods are readily integrated into diverse deep learning architectures without necessitating complex sampling schemes [46–48]. Empirical investigations have further substantiated that class-balanced loss techniques are often a superior choice for SSL, especially in the context of pseudo-label generation [43]. Nevertheless, when the imbalance ratio differs significantly between labeled and unlabeled datasets, class-balanced loss methods may exhibit reduced efficacy in pseudo-label generation. In this paper, we advocate for a class re-balancing approach within a prototypical network framework, as prototypical networks are inherently well-suited to address class imbalance challenges [42]. This architectural approach initially projects class-imbalanced features onto class prototypes within a latent space and subsequently learns class assignments for each prototype while concurrently re-balancing class representations.

DCISS integrates SSL and class-imbalanced learning within a cohesive framework by incorporating a prototypical network component. The architectural blueprint of DCISS is illustrated in Fig. 4.1. Given a partially labeled dataset characterized by class imbalance, we employ a feature encoder $\Psi$, instantiated as a CNN, to project the input data into a deep latent space. The rationale for selecting a CNN architecture stems from its established efficacy in feature extraction and rapid convergence properties. Furthermore, we have augmented the CNN feature encoder with a channel calibration attention module to further enhance feature extraction capabilities and accelerate model convergence. Upon feature projection, the prototypical network calculates a representative prototype for each class (icing or non-icing) based on the labeled data. Following the insights of [38], the prototype derived from unlabeled samples is instrumental in model training enhancement. Consequently, we iteratively refine the prototype of each class by incorporating information from the unlabeled sample prototype (as depicted in the bottom panel of Fig. 4.1). However, it is crucial to emphasize that the integration of unlabeled samples during training is solely for the purpose of model optimization and enrichment of training data. In the inference phase, each test sample is mapped into the deep latent space via the feature extraction function, $\Psi(\cdot)$, and subsequently classified by the trained discriminative classifier.

### *4.2.3 Feature Encoder Architecture*

The feature encoder is implemented as a convolutional neural network (CNN), comprising three serially connected blocks as depicted in Fig. 4.1. Each block is structurally composed of a one-dimensional convolution layer (Conv1D), a channel calibration attention layer (Ch_Attn), a normalization layer (Norm1D), and a non-linear activation layer (ReLU). Within each block, the Ch_Attn layer is integrated to enhance feature learning efficacy and expedite convergence. For an input $X_{raw} \in \mathbb{R}^{d \times T}$, where $d$ and $T$ represent the dimensionality and temporal window size of the input samples, respectively, the output transformations within each CNN block can be mathematically represented as follows:

$$\begin{aligned} X_c &= Conv1D(X_{raw}), \\ X_{Attn} &= Ch\_Attn(X_c), \\ X_{AttnCNN} &= ReLU(Norm1D(X_{Attn})), \end{aligned} \tag{4.1}$$

where $X_c$, $X_{Attn}$, and $X_{AttnCNN}$ denote the output feature maps from the Conv1D layer, attention layer, and ReLU layer, respectively. All three output tensors maintain a consistent shape of $\mathbb{R}^{F \times T}$, where $F$ signifies the number of filters employed in the Conv1D layer.

The operational process of the Ch_Attn module is illustrated in Fig. 4.2. The input to Ch_Attn is the output feature map from the preceding Conv1D layer, possessing a shape of $\mathbb{R}^{F \times T}$ . As previously mentioned, $F$ represents the count of 1D CNN filters, each corresponding to a distinct feature channel. Conventional CNN architectures typically treat all feature channels uniformly, which can inadvertently lead to the loss of salient information [17]. To address this limitation, we incorporate the Ch_Attn layer to selectively emphasize the most informative feature channels.

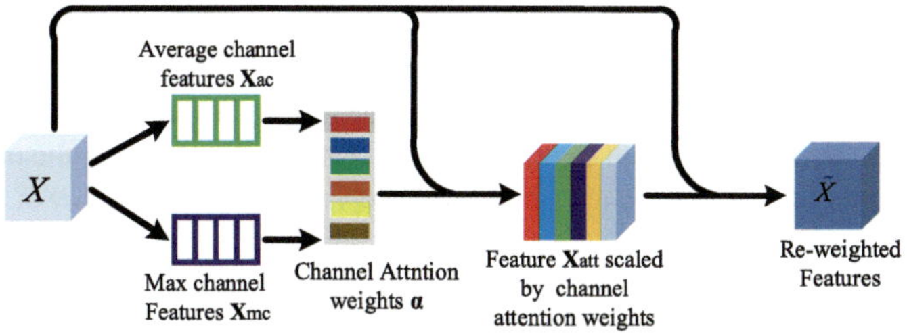


**Fig. 4.2** Detailed illustration of the channel calibration attention module

Let us assume that the output of the Conv1D layer is the feature map, $X_c = [x_1, x_2, ..., x_F]$ where $x_k \in \mathbb{R}^{T\times 1}$ represents the $k$-th channel. To discern and prioritize informative channels, we apply global average pooling and global max pooling operations to derive distinctive feature vectors, $X_{ac}$ and $X_{mc}$, respectively (refer to Fig. 4.2). Both resultant feature vectors have a shape of $\mathbb{R}^{F\times 1}$. Subsequently, these vectors are fed into two multi-layer perceptrons (MLPs) with shared parametric weights (not explicitly shown in Fig. 4.2) to compute channel-wise attention weights $\alpha$, formally defined as

$$\begin{aligned} \alpha &= \sigma(\boldsymbol{W_2}(\boldsymbol{W_1}(X_{ac})) + \boldsymbol{W_2}(\boldsymbol{W_1}(X_{mc}))), \\ \boldsymbol{X_{att}} &= \alpha \otimes \boldsymbol{X}, \end{aligned} \tag{4.2}$$

where $\boldsymbol{X}$ is the input to the feature encoder, $\alpha$ denotes the computed attention weights, $\boldsymbol{X_{att}}$ represents the feature map scaled by the attention weights, and $\otimes$ signifies element-wise multiplication. $\boldsymbol{W_1} \in \mathbb{R}^{F/r\times F}$ and $\boldsymbol{W_2} \in \mathbb{R}^{C\times F/r}$ are the weight matrices of the first and second MLPs, respectively, where $r$ is a reduction ratio. Inspired by the demonstrated efficacy of residual blocks [49], we integrate channel attention with a residual connection mechanism, termed residual channel attention. That is, in Fig. 4.2, we have $\widetilde{\boldsymbol{X}} = \boldsymbol{X} + \boldsymbol{X_{att}}$, where $\widetilde{\boldsymbol{X}}$ represents the attention-weighted feature map, $\boldsymbol{X}$ is the input feature map to the encoder, and $\boldsymbol{X_{att}}$ is the feature map scaled by the attention weight vector, $\alpha$.

### *4.2.4 Prototypical Learning Mechanism*

In prototypical networks, a prototype is defined as a representative vector for each class. In this research, the prototype for labeled samples is computed as the centroid of all feature vectors belonging to a specific class in the deep latent space, formally defined as

$$C_l^i = \frac{1}{n_l^i}\sum_{k=1}^{n_l^i} X_l^{i,k}, \tag{4.3}$$

where $C_l^i \in \mathbb{R}^{1\times H}$ is the computed prototype for class $i \in [1, \cdots, n_c]$ in the labeled dataset (for simplicity, we use natural numbers to index classes), $n_c$ is the total number of classes, $n_l^i$ is the number of labeled instances belonging to class $i$ in the deep latent space, and $X_l$ denotes the set of feature vectors for labeled samples in the deep latent space. For unlabeled samples, a global prototype is computed as follows:

$$C_u = \frac{1}{n_u}\sum_{k=1}^{n_u} X_u^k, \tag{4.4}$$

where $C_u \in \mathbb{R}^{1\times H}$ is the computed prototype for the ensemble of unlabeled samples, $n_u$ is the total number of unlabeled samples in the deep latent space, and $X_u$ represents the set of extracted unlabeled feature vectors in the deep latent space.

In this study, we compute a single global prototype for all unlabeled samples. Subsequently, we refine each class prototype derived from the labeled samples by incorporating information from the unlabeled prototype, defined as follows:

$$C_{new}^i = \frac{\sum_{k=1}^{n_l^i} X_l^{i,k} + \sum_{k=1}^{n_u} X_u^k}{2(n_l^i + n_u)}. \tag{4.5}$$

Given that the unlabeled prototype encapsulates information from both classes (icing and non-icing), its integration effectively enhances the information entropy of the class prototypes derived from labeled data. Consequently, we update both labeled prototypes based on the unlabeled prototype, weighted by the relative distances between the unlabeled and labeled prototypes. This proposed prototype update mechanism is computationally efficient yet demonstrably effective, as empirically validated by the experimental results detailed in Sect. 4.3.

After obtaining the class prototypes, the class membership probability for a given time series can be estimated by quantifying its proximity to each prototype in the deep latent space. For an input sample $x \in \mathbb{R}^d$, its projected feature vector in the deep latent space is $\Psi(x)$. The probability of sample $x$ belonging to class $k$ is computed using a Softmax function based on the Euclidean distances to the class prototypes in the deep latent space:

$$p(y = k|x) = \frac{exp(-dist(\Psi(x), C_{new}^k))}{\sum_i exp(-dist(\Psi(x), C_{new}^i))}, \tag{4.6}$$

where $dist$ is a distance metric between two vectors. Here, we employ the Euclidean distance metric due to its empirically demonstrated efficacy in prototypical networks [50]. Subsequently, we utilize the categorical cross-entropy loss function to guide model training, defined as follows:

$$\mathcal{L} = -\sum_{(x_i, y_i)} log(p(y = y_i|x_i, C_{new}^k)), \tag{4.7}$$

where $(x_i, y_i)$ represents labeled samples in the training dataset. $C^k_{new}$ is the updated prototype corresponding to the class label $y_i$ of sample $(x_i, y_i)$. Model parameters are optimized by minimizing the average loss over the training dataset, iteratively proceeding through training epochs, and performing gradient descent updates in each iteration. In this paper, we employ the back-propagation algorithm for gradient computation and the Adam optimizer for loss function minimization, owing to its computational efficiency and reduced memory footprint. Algorithms 1 and 2 provide a more granular algorithmic specification of the learning and inference processes, respectively.

**Algorithm 1** Learning process of DCISS

**Input** : Number of epochs: $T$
**Output**: Well-trained feature encoder: $\Psi$, Prototypes of classes: $C_{new}$
**Data** : Labeled samples: $S_l$ and the corresponding labels: $Y_l$; Unlabeled samples: $S_u$
**for** $iter \leftarrow 1$ *to* $T$ **do**
  $X_l \leftarrow \Psi(S_l)$ ; `/* Extract features of labeled samples in latent space */`
  $X_u \leftarrow \Psi(S_u)$ ; `/* Extract features of unlabeled samples in latent space */`
  **for** $i \leftarrow 1$ *to* $n_c$ **do**
    $idx \leftarrow$ Get indices of class $i$ in $Y_l$
    $C^i_l \leftarrow \frac{1}{n^i_l}\sum_{k=1}^{n^i_l} X^{i,k}_l(idx)$ ; `/* Eq.(4.3) */`
  **end**
  $C_u \leftarrow \frac{1}{n_u}\sum_{k=1}^{n_u} X^k_u$ ; `/* Eq.(4.4) */`
  **for** $i \leftarrow 1$ *to* $n_c$ **do**
    $C^i_{new} \leftarrow \frac{\sum_{k=1}^{n^i_l} X^{i,k}_l + \sum_{k=1}^{n_u} X^k_u}{2(n^i_l + n_u)}$ ; `/* Eq.(4.5) */`
  **end**
  $p(y = k|x) \leftarrow \frac{exp(-dist(\Psi(x), C^k_{new}))}{\sum_i exp(-dist(\Psi(x), C^i_{new}))}$ ; `/* Eq.(4.6) */`
  $\mathcal{L} = -\sum_{(x_i, y_i)} log(p(y = y_i|x_i, C^k_{new}))$ ; `/* Eq.(4.7) */`
  $minimize\ \mathcal{L}$ ; `/* Minimize loss function using Adam optimizer and backpropagation */`
**end**

## 4.3 Experiments

### *4.3.1 Experimental Configuration*

#### (1) Implementation Details

The deep learning algorithms were implemented using PyTorch v1.6.0 and Python v3.8.0. All experiments were conducted on a server equipped with NVIDIA Titan V GPUs. Hyperparameter optimization was deliberately omitted; instead, we adopted

**Algorithm 2** Testing process of DCISS

**Input** : Feature encoder: $\Psi$; Prototypes of classes: $C_{new}$
**Output**: Set of labels: $L$
**Data** : Testing dataset: $V$

```
L ← {}
for each v ∈ V do
    X_v ← Ψ(v) ; /* Extract features of test sample v in latent space */
    p ← exp(−dist(X_v, C_new)) / Σ_i exp(−dist(X_v, C^i_new)) ; /* Calculate probability distribution over classes */
    l ← argmax(p) ; /* Assign class label with maximum probability */
    L ← L ∪ {l}
end
```

training parameters consistent with those used in the baseline models to ensure comparability.

**(2) Datasets**

The datasets employed in these experiments were provided by Goldwind Inc., a prominent wind energy corporation globally. Data acquisition was performed using the SCADA system from two wind turbines located in Inner Mongolia, China. One turbine's data spanned 306 hours, while the other encompassed 696 hours, both sampled at a 7-second resolution. The sensor suite comprised over one hundred distinct signals (a subset is illustrated in Fig. 4.3). Wind turbine domain experts meticulously annotated the entire dataset, identifying 26 variables pertinent to blade icing, as detailed in Table 4.1. We aggregated the datasets from both wind turbines and partitioned them into 70% for training (with approximately 10% labeled data and 60% unlabeled data) and 30% for testing. It is important to note that the training dataset contained both labeled and unlabeled instances, exhibiting class imbalance, whereas the testing dataset was intentionally balanced to provide a robust evaluation.

To rigorously evaluate the impact of data imbalance, we designed three distinct experimental scenarios, characterized by varying imbalance ratios: $\rho_u = \rho_l$, $\rho_u =$

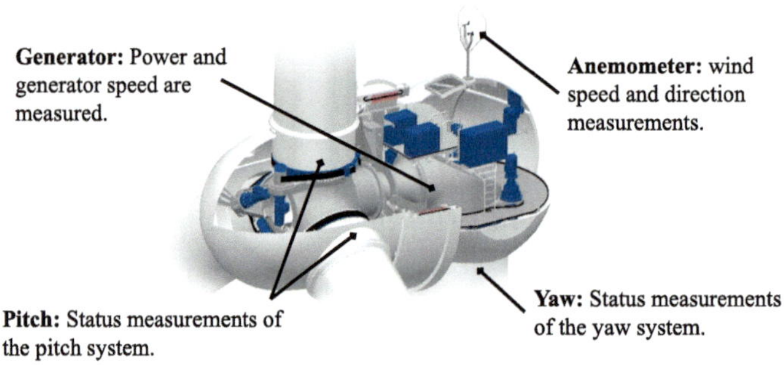


**Fig. 4.3** Illustrative depiction of sensor placements within a wind turbine system

**Table 4.1** Wind Turbine SCADA data parameters

| Index | Monitored variable | Index | Monitored variable |
|---|---|---|---|
| 1 | Wind Velocity | 14 | Pitch Motor 1 Temperature |
| 2 | Generator Rotational Speed | 15 | Pitch Motor 2 Temperature |
| 3 | Active Electrical Power Output | 16 | Pitch Motor 3 Temperature |
| 4 | Wind Direction (Compass Heading) | 17 | Horizontal Acceleration (Turbine Structure) |
| 5 | 25-Second Averaged Wind Direction Angle | 18 | Vertical Acceleration (Turbine Structure) |
| 6 | Yaw System Position | 19 | Ambient Temperature |
| 7 | Yawing Rate | 20 | Nacelle Internal Temperature |
| 8 | Pitch Angle - Blade 1 | 21 | Pitch 1 Switching Element Temperature |
| 9 | Pitch Angle - Blade 2 | 22 | Pitch 2 Switching Element Temperature |
| 10 | Pitch Angle - Blade 3 | 23 | Pitch 3 Switching Element Temperature |
| 11 | Pitch Actuator 1 Speed | 24 | Pitch 1 Charger DC Power Consumption |
| 12 | Pitch Actuator 2 Speed | 25 | Pitch 2 Charger DC Power Consumption |
| 13 | Pitch Actuator 3 Speed | 26 | Pitch 3 Charger DC Power Consumption |

$2\rho_l$, and $\rho_u = 1$. Here, $\rho$ represents the imbalance ratio between the normal and icing classes ($u$ and $l$ subscripts denote unlabeled and labeled data, respectively), formally defined as

$$\rho = \frac{N_{normal}}{N_{icing}}, \tag{4.8}$$

where $N_{normal}$ and $N_{icing}$ are the counts of samples labeled as normal and icing, respectively.

**(3) Performance Metrics**

Given the inherent class imbalance in the dataset, conventional accuracy metrics are not optimally suited for performance evaluation. Consequently, we employed **Precision**, **Recall**, **F1-Score**, and the **Matthews Correlation Coefficient** (MCC) as evaluation metrics, defined as follows:

$$Precision = \frac{TP}{TP + FP}, \tag{4.9}$$

$$Recall = \frac{TP}{TP + FN}, \tag{4.10}$$

$$F1 = \frac{2 \times Precision \times Recall}{Precision + Recall}, \tag{4.11}$$

$$MCC = \frac{TP \cdot TN - FP \cdot FN}{\sqrt{(TP + FP)(TP + FN)(TN + FP)(TN + FN)}}, \tag{4.12}$$

where $TP$, $FP$, $FN$, and $TN$ represent true positives, false positives, false negatives, and true negatives, respectively. Given the critical importance of maximizing the detection of icing events, recall performance is of paramount significance. The F1-score provides a balanced measure of precision and recall, suitable for assessing overall performance. MCC is a specialized binary classification metric particularly robust for evaluating imbalanced classification scenarios.

### *4.3.2 Comparative Evaluation with Anomaly Detection and Semi-supervised Learning Baselines*

In this investigation, we benchmarked our proposed DCISS model against five established baselines for anomaly detection and SSL:

- **IsForest** (Isolation Forest): We adopted the parameter settings recommended by [23], configuring the number of trees to 100 and the sub-sampling size to 256.
- **OCSVM** (One-Class SVM): An RBF kernel was employed. We systematically evaluated performance across RBF scale parameters {0.01, 0.05, 0.1, 0.2, 0.5} and report the optimal result achieved.
- **LOF** (Local Outlier Factor): This is an outlier detection algorithm that quantifies the local outlier factor to reflect the degree of anomaly of test instances [24].
- **TapNet**: This is a state-of-the-art semi-supervised algorithm for TSC. TapNet incorporates a feature extractor comprising an LSTM layer and stacked CNN layers. An attention-based prototypical network is implemented for SSL [38].
- **ImbalancedSSL** (Imbalanced Semi-Supervised Learning): This algorithm addresses class-imbalanced SSL, utilizing ResNet as a feature extractor [43]. It represents a state-of-the-art methodology for class-imbalanced semi-supervised learning. We substituted the ResNet architecture with our proposed feature encoder to facilitate a direct performance comparison.

The temporal window size for sensor data was set to 128 samples (equivalent to 15 min). The number of filters in the feature extractor was configured as {128, 256, 128} with kernel sizes of {7, 5, 3}. These parameter settings were consistently applied across all subsequent experiments without further elaboration. Each experiment was replicated three times, and the best observed outcome is reported. The results are presented in Table 4.2, with the highest performance metrics highlighted in bold. As evident from the table, our proposed DCISS model achieves superior performance with respect to F1-score and MCC across all imbalance ratio scenarios. In contrast, the three classical unsupervised anomaly detection algorithms (IsForest,

**Table 4.2** Performance comparison against anomaly detection and semi-supervised learning baselines

| Method | $\rho_u = \rho_l$ | | | | $\rho_u = 2\rho_l$ | | | | $\rho_u = 1$ | | | |
|---|---|---|---|---|---|---|---|---|---|---|---|---|
| | Precision | Recall | F1 | MCC | Precision | Recall | F1 | MCC | Precision | Recall | F1 | MCC |
| IsForest | 51.7 | 97.2 | 53.2 | 13.2 | 52.6 | 97.7 | 54.8 | 18.7 | 52.9 | 95.6 | 55.3 | 17.8 |
| LOF | 50.2 | **99.5** | 50.3 | 2.55 | 50.0 | **99.4** | 50.1 | 1.24 | 50.3 | **99.6** | 50.5 | 5.04 |
| OCSVM | 51.2 | 94.9 | 52.2 | 8.26 | 51.7 | 94.7 | 53.2 | 11.5 | 51.9 | 94.1 | 53.4 | 11.6 |
| TapNet | 92.0 | 46.9 | 71.4 | 49.1 | 88.7 | 31.0 | 63.5 | 35.6 | 88.2 | 41.7 | 68.1 | 42.7 |
| ImbalancedSSL | **96.1** | 50.3 | 74.1 | 54.9 | **96.5** | 43.4 | 70.9 | 50.1 | **96.3** | 51.3 | 74.7 | 55.8 |
| **DCISS (Ours)** | 91.9 | 86.1 | **89.2** | **78.6** | 90.7 | 89.8 | **90.3** | **80.5** | 92.8 | 77.7 | **85.8** | **72.7** |

LOF, OCSVM) exhibit the lowest performance overall; surprisingly, however, they attain the highest recall values. In comparison to TapNet, our model demonstrates a substantial improvement in MCC of 60.0, 126.1, and 70.3% for the three imbalance ratio conditions. The corresponding F1-score improvements are 24.9, 42.2, and 26.0%. ImbalancedSSL, a state-of-the-art method for class-imbalanced SSL, exhibits performance inferior to our DCISS model in terms of F1-score and MCC, although it demonstrates the highest precision. Regarding the influence of varying imbalance ratios, our model achieves peak performance when $\rho_u = 2\rho_l$.

### *4.3.3 Evaluation of Class-Imbalanced Learning Efficacy*

We conducted a comparative assessment of the proposed DCISS model against five established class-imbalanced learning algorithms. In this evaluation, we considered labeled sample proportions of 10, 30, and 50%, while maintaining a consistent imbalance ratio (see Eq. (4.8)) of 10. The five comparative algorithms are detailed below:

- **Focalloss**: This is a widely adopted loss function for class-imbalanced learning. In this study, we set $\gamma = 1$ and employed class weights as recommended in [46].
- **GaussianAffinity**: This algorithm represents a flexible loss function tailored for class-imbalanced learning. Grounded in Euclidean space, affinity is defined using Gaussian similarity within Bregman divergence. GaussianAffinity is applied for feature space clustering and max-margin classification based on prototypes [51].
- **Class-balance**: This algorithm introduces the concept of the effective number of samples, explicitly accounting for data overlap. The class-balanced loss is calculated based on the effective number of samples per class [47].
- **LDAM** (Label-Distribution-Aware Margin): This algorithm is designed as a replacement for the standard cross-entropy objective during training and can be integrated with prior class-imbalanced training strategies, such as re-weighting or resampling [48].
- **WCE** (Weighted Cross Entropy): This algorithm utilizes a classical loss function for class-imbalanced learning, assigning a distinct weight to each class to address imbalance [43].

In this experimental phase, we evaluated the class re-balancing capacity of our proposed model in isolation, focusing solely on the imbalance correction component. The results are presented in Table 4.3, demonstrating that our DCISS model achieves the highest performance in terms of precision, F1 and MCC. LDAM ranks as the second-best performing method, while GaussianAffinity exhibits the lowest performance. Our model achieves MCC improvements of 82.6, 59.9, and 42.1% over LDAM for the three labeled sample proportions, and F1-score improvements of 18.9, 19.4, and 14.5%. Interestingly, WCE slightly outperforms Focalloss, potentially attributable to the fact that we employed default hyperparameters without fine-tuning. The results also indicate a general trend of performance enhancement with increasing

**Table 4.3** Performance comparison across varying proportions of labeled samples with an imbalance ratio of 10

| Method | 10% | | | | 30% | | | | 50% | | | |
|---|---|---|---|---|---|---|---|---|---|---|---|---|
| | Precision | Recall | F1 | MCC | Precision | Recall | F1 | MCC | Precision | Recall | F1 | MCC |
| Focalloss | 59.4 | 70.5 | 61.2 | 22.9 | 66.9 | 74.6 | 68.8 | 38.0 | 67.6 | 82.8 | 71.6 | 44.3 |
| GaussianAffinity | 54.0 | 77.4 | 55.7 | 12.6 | 54.8 | 67.8 | 56.0 | 13.3 | 54.1 | 77.8 | 55.9 | 13.1 |
| Class-balance | 55.1 | **83.3** | 57.7 | 21.6 | 55.3 | **94.7** | 59.1 | 26.0 | 55.1 | **92.9** | 58.6 | 25.9 |
| LDAM | 72.7 | 51.1 | 66.0 | 33.4 | 76.8 | 71.4 | 74.9 | 49.9 | 81.7 | 70.6 | 77.4 | 55.4 |
| WCE | 65.7 | 55.9 | 63.3 | 27.0 | 68.2 | 76.4 | 70.4 | 41.1 | 76.9 | 68.4 | 73.9 | 48.4 |
| **DCISS (Ours)** | **94.4** | 60.5 | **78.5** | **61.0** | **97.0** | 81.3 | **89.4** | **79.8** | **97.9** | 79.0 | **88.6** | **78.7** |

proportions of labeled samples. Finally, in terms of recall, the Class-balance method demonstrates the highest performance.

### *4.3.4 Feature Encoder Architecture Evaluation*

To rigorously evaluate the proposed feature encoder, we compared its performance against five alternative neural network architectures as baselines, described as follows:

- **MLP (Multilayer Perceptron)**: The MLP architecture comprised three fully connected (FC) layers, each with 500 hidden nodes, and dropout regularization layers interspersed between FC layers.
- **LSTM (Long Short-Term Memory Network)**: We systematically varied the number of hidden nodes in the LSTM layer across {8, 16, 32, 64}, evaluated performance for each configuration, and reported the best result obtained.
- **MLSTM-FCN (Multilayer LSTM - Fully Convolutional Network)**: This deep learning model integrates an LSTM layer and an FCN layer, incorporating a Squeeze-and-Excitation (SE) module. We utilized the parameter settings recommended in [32].
- **ResNet (Residual Network)**: We adopted the identical ResNet architecture and parameters as specified in [52].
- **DenseNet (Densely Connected Convolutional Network)**: This model has been reported to achieve state-of-the-art performance in time-series classification [17]. We implemented the same network structure but deliberately removed all attention layers to isolate the effect of network topology.

The comparative results are presented in Table 4.4. The results indicate that our proposed feature encoder achieves the highest performance across the three test cases (with the exception of F1-score in the $\rho_u = 1$ scenario, where MLSTM_FCN achieves comparable performance). MLSTM_FCN consistently ranks second in performance, except in the $\rho_u = \rho_l$ test case, where its performance is marginally inferior to DenseNet, possibly due to its inherent capacity to concurrently learn spatial and temporal features. ResNet and DenseNet exhibit nearly identical performance across all three test conditions. LSTM demonstrates the lowest performance, even underperforming MLP in all three scenarios. In terms of MCC, our proposed model shows performance gains of 28.2, 33.7, and 0.28% over the second-ranked method for the three test cases, respectively. For F1-score, our model achieves absolute improvements of 9.9 and 11.7% in the $\rho_u = \rho_l$ and $\rho_u = 2\rho_l$ test cases, respectively. Another noteworthy observation is that performance does not consistently improve with increasing network complexity. For instance, DenseNet, despite possessing the most complex architecture among the evaluated models, does not achieve the highest performance. This observation underscores the importance of selecting an appropriate model architecture, which can be more critical than simply increasing the depth or complexity of the network.

**Table 4.4** Feature encoder performance comparison

| Method | $\rho_u = \rho_l$ | | | | $\rho_u = 2\rho_l$ | | | | $\rho_u = 1$ | | | |
|---|---|---|---|---|---|---|---|---|---|---|---|---|
| | Precision | Recall | F1 | MCC | Precision | Recall | F1 | MCC | Precision | Recall | F1 | MCC |
| MLP | 62.6 | 62.1 | 62.5 | 24.9 | 59.7 | 64.3 | 60.4 | 20.9 | 58.3 | 68.2 | 59.7 | 19.7 |
| LSTM | 57.3 | 56.8 | 57.2 | 14.4 | 57.3 | 62.5 | 58.0 | 16.0 | 54.6 | 55.6 | 54.7 | 9.3 |
| ResNet | 90.7 | 65.4 | 79.3 | 61.1 | 92.6 | 54.6 | 75.1 | 55.1 | 89.8 | 72.0 | 81.9 | 65.2 |
| DenseNet | 89.7 | 67.0 | 79.7 | 61.3 | 92.1 | 55.6 | 75.4 | 55.4 | 91.0 | 71.9 | 82.4 | 66.2 |
| MLSTM_FCN | 89.7 | 66.2 | 79.3 | 60.7 | 91.6 | 63.0 | 78.6 | 60.2 | 92.2 | **78.2** | **85.8** | 72.5 |
| **DCISS (Ours)** | **91.9** | **86.1** | **89.2** | **78.6** | 90.7 | **89.8** | **90.3** | **80.5** | **92.8** | 77.7 | **85.8** | **72.7** |

### 4.3.5 Sensitivity and Ablation Analyses

We conducted sensitivity and ablation analyses to comprehensively evaluate the effectiveness and robustness of the proposed DCISS model.

#### 4.3.5.1 Influence of Imbalance Factor

We assessed the model's performance under varying imbalance ratios (see Eq. (4.8)): $\rho_l = 10, 20$ and 50; for each $\rho_l$, we correspondingly adjusted $\rho_u$. The results are presented in Table 4.5. The results indicate that the proposed model exhibits optimal performance when the imbalance ratio is configured to 10. Furthermore, a discernible performance degradation is observed as the imbalance ratio increases from 10 to 50.

#### 4.3.5.2 Influence of Labeled Sample Quantity

To investigate the impact of labeled sample quantity, we varied the proportion of labeled samples within the training dataset from 10 to 50%. For each labeled data proportion, we conducted three comparative experiments, examining the influence of the relationship between imbalance ratios and labeled sample size on classification performance. Table 4.6 presents the results, indicating that for $\rho_u = \rho_l$ and $\rho_u = 2\rho_l$ scenarios, the model achieves maximal performance when the labeled sample proportion is 50%. Interestingly, in the $\rho_u = 1$ case, model performance at 50% labeled samples is marginally lower than at 30% labeled samples. These findings suggest a complex interplay between labeled data quantity, imbalance ratio, and overall classification efficacy.

#### 4.3.5.3 Impact of Temporal Window Size

We evaluated the impact of varying temporal window sizes for data segmentation during training, considering sizes of 32, 64, 128, and 256 samples. The imbalance ratios were set to $\rho_u = 20$ and $\rho_l = 10$, and the labeled sample proportion in the training set was maintained at 10%. The results, depicted in Fig. 4.4, demonstrate that the DCISS model achieves optimal performance in terms of F1-score and MCC when the temporal window size is set to 128 samples. As the window size increases from 32 to 128, MCC and F1-scores also increase, but a slight performance reduction is observed at a window size of 256. This suggests that while larger window sizes capture more temporal context, excessively large windows may introduce noise or redundancy, thus impacting performance. Selecting an appropriate temporal window size for data segmentation is therefore crucial for optimal model training; in this study, a window size of 128 samples was empirically determined to be optimal.

**Table 4.5** Influence of imbalance ratio on performance

| Imbalance Factor | $\rho_u = \rho_l$ | | | | $\rho_u = 2\rho_l$ | | | | $\rho_u = 1$ | | | |
|---|---|---|---|---|---|---|---|---|---|---|---|---|
| | Precision | Recall | F1 | MCC | Precision | Recall | F1 | MCC | Precision | Recall | F1 | MCC |
| $\rho_l = 10$ | **91.9** | **86.1** | **89.2** | **78.6** | 90.7 | **89.8** | **90.3** | **80.5** | **92.8** | **77.7** | **85.8** | **72.7** |
| $\rho_l = 20$ | 88.3 | 64.2 | 77.9 | 58.7 | **91.4** | 61.5 | 77.9 | 59.0 | 85.5 | 64.6 | 76.8 | 57.8 |
| $\rho_l = 50$ | 83.6 | 52.2 | 71.0 | 47.4 | 67.1 | 82.5 | 71.0 | 47.7 | 72.0 | 76.7 | 73.4 | 47.0 |

**Table 4.6** Influence of labeled sample quantity on performance

| Number of Labeled Samples | $\rho_u = \rho_l$ | | | | $\rho_u = 2\rho_l$ | | | | $\rho_u = 1$ | | | |
|---|---|---|---|---|---|---|---|---|---|---|---|---|
| | Precision | Recall | F1 | MCC | Precision | Recall | F1 | MCC | Precision | Recall | F1 | MCC |
| 10% | 91.9 | 86.1 | 89.2 | 78.6 | 90.7 | **89.8** | 90.3 | 80.5 | 92.8 | 77.7 | 85.8 | 72.7 |
| 30% | 95.5 | 87.3 | 91.6 | 83.5 | 96.1 | 86.9 | 91.7 | 83.7 | 96.7 | **83.9** | **90.5** | **81.8** |
| 50% | **97.8** | **86.8** | **92.4** | **85.4** | **97.8** | 85.6 | **91.8** | **84.3** | **97.7** | 81.9 | 90.0 | 81.0 |

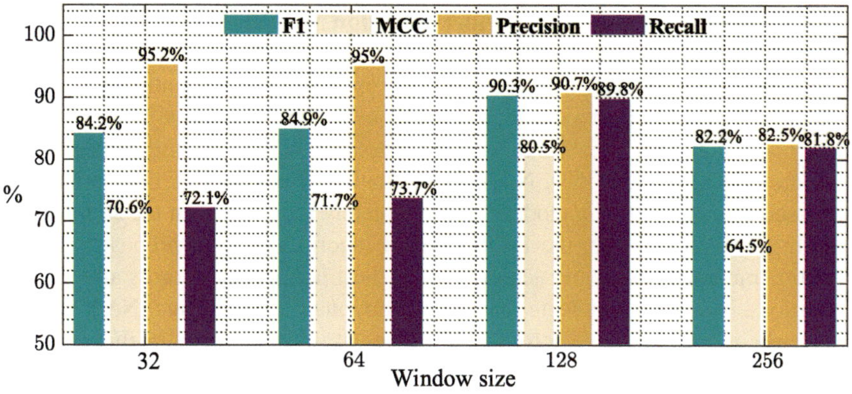


**Fig. 4.4** Influence of different window sizes on performance

#### 4.3.5.4 Impact of Convolutional Filter Size

The filter size is a critical hyperparameter in the feature encoder architecture. We systematically investigated its influence by varying the filter size across values of 16, 32, 64, and 128. The imbalance ratios were set to $\rho_u = 20$ and $\rho_l = 10$. The results, presented in Fig. 4.5, reveal that optimal performance is attained when the filter size is set to 128. As the filter size is increased from 16 to 128, MCC and F1-scores exhibit a corresponding increase. These results underscore the sensitivity of the proposed model's performance to the convolutional filter size. Consequently, judicious selection of the filter size is paramount in real-world applications to maximize detection accuracy.

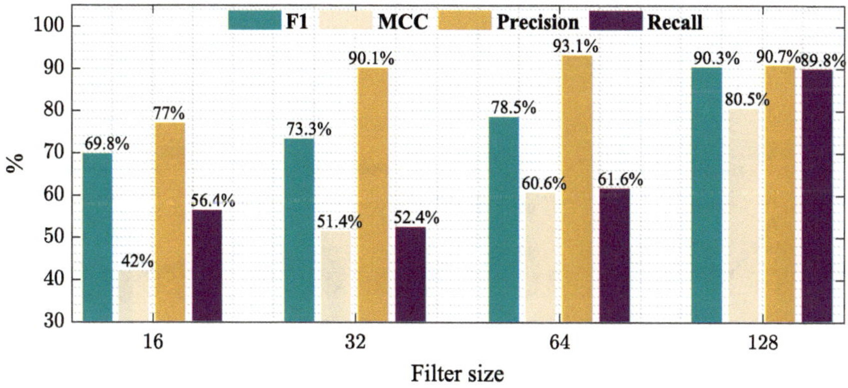


**Fig. 4.5** Influence of different filter sizes on performance

#### 4.3.5.5 Effectiveness of the Channel Attention Module

To rigorously assess the effectiveness of the proposed channel attention module, we compared the following model variants: **SE** (Squeeze-and-Excitation): In this variant, the proposed channel attention module was replaced with the classic Squeeze-and-Excitation (SE) module [53]. **NAN** (No Attention Network): In this variant, the proposed channel attention module was completely removed from the architecture. As illustrated in Fig. 4.6, the DCISS model incorporating the proposed channel attention module consistently achieves the highest performance across all three test conditions. The results indicate that the performance of the **SE** and **NAN** variants are nearly indistinguishable across all test cases. It is also noteworthy that precision metrics are largely comparable across the three models, but recall values for the **SE** and **NAN** variants are demonstrably lower than for our proposed model. Regarding F1-score and MCC, the three models exhibit similar performance in the $\rho_u = 1$ test case. However, for F1-score, our proposed model outperforms **NAN** by 9.5 and 12.3% in the $\rho_u = \rho_l$ and $\rho_u = 2\rho_l$ tests, respectively. Similarly, for MCC, our model surpasses **NAN** by 17.2 and 21.6% in these respective test scenarios.

### 4.3.6 *Online Icing Detection Evaluation*

To evaluate the real-time blade icing detection capability of the DCISS model in a practical wind turbine setting, we implemented an online detection scheme. In this scheme, the model deployed for online detection was derived from the offline model training process. A data buffer, with a temporal length equivalent to the window size used during model training, was employed to accumulate incoming online data streams. Upon buffer saturation, the scheme performs icing detection by computing a probability value. We incorporated an ensemble algorithm, specifically majority voting [12], into this scheme to enhance robustness and mitigate random error propagation.

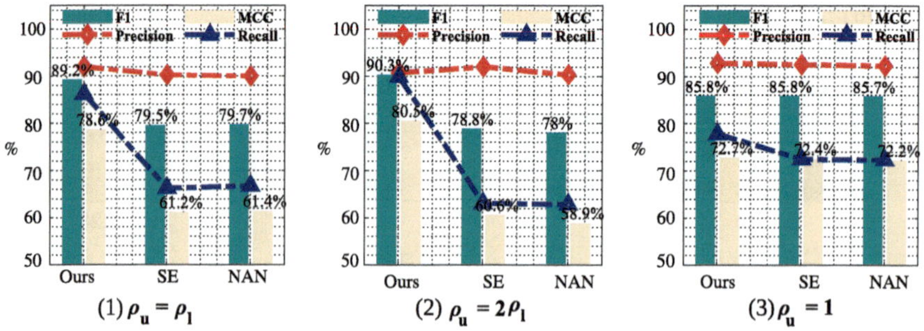


**Fig. 4.6** Ablation study for evaluating the channel attention module

The online detection evaluation was conducted using a DCISS model trained on data from the two wind turbines described in Sect. 4.3.1. The trained model was then applied to a third, independent wind turbine dataset, comprising continuous logging data spanning approximately 330 hours. To further assess the robustness of our model, we compared its online detection performance against a supervised learning model. The DCISS model was trained using 10% labeled data and 60% unlabeled data, sourced from the combined datasets of the initial two wind turbines. The supervised learning model was trained using 50% labeled samples, also derived from the combined datasets of the two turbines. Blade icing condition predictions were generated approximately every 4 min, over a continuous 23-h operational period. The results are visualized in Fig. 4.7, where the magenta-shaded region represents the periods of blade icing as annotated by a turbine engineer. The green dashed horizontal line denotes the conventional threshold of 0.5 for binary icing detection. Within the icing period (magenta region), both methods achieve near-perfect accuracy in icing prediction. The supervised method issues an icing alarm at approximately 3,500 s, whereas the semi-supervised DCISS method triggers an alarm at approximately 7,200 seconds. Thus, the semi-supervised DCISS model demonstrates superior performance in accurately predicting the normal operating state of the turbine, exhibiting fewer false alarms. Although the DCISS model predicts icing onset at a slightly later time point compared to the supervised method, the temporal delay is still sufficient to allow turbine engineers to proactively initiate de-icing measures (i.e., activate a de-icing system) prior to significant ice accretion.

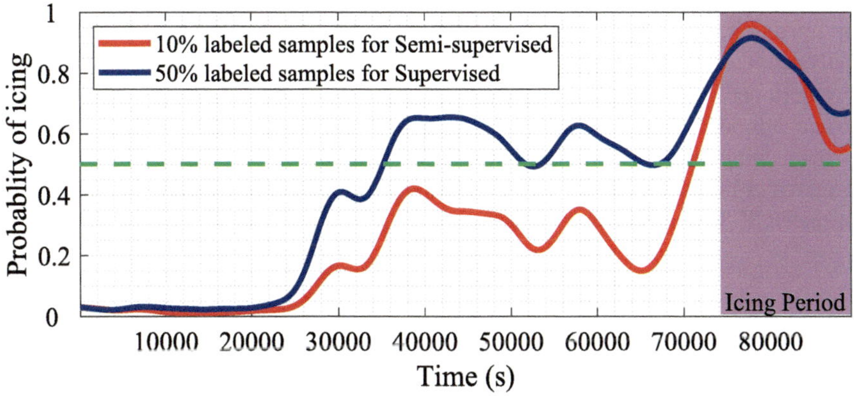


**Fig. 4.7** Evaluation of DCISS model performance in online icing detection scenario

## 4.4 Discussion

In this research endeavor, we introduced a novel prototypical network architecture that unifies class re-balancing and semi-supervised learning methodologies within a single, cohesive framework for class estimation on unlabeled data. While the empirical evaluations have validated its efficacy and superior performance in wind turbine blade icing detection, two aspects warrant further discussion and consideration. Firstly, experimental results indicated that our DCISS model achieves optimal performance under an imbalance ratio of 10, with a subsequent performance decrement observed at imbalance ratios of 20 and 50. The underlying mechanisms responsible for this performance variation across different imbalance ratios remain an open question. This phenomenon may be intrinsically linked to the network architecture design and merits further in-depth investigation in future research. Secondly, the current DCISS model implementation is not inherently designed for quantifying the severity of blade icing in wind turbines. This limitation stems from the binary nature of the training data, which only distinguishes between two discrete states: normal (non-icing) and icing. Consequently, extending the model's capability to detect and classify continuous icing severity states, including intermediate states between normal and fully iced conditions, represents a significant direction for future model refinement. A potential pragmatic workaround within the current implementation, as demonstrated in the online testing scenario, involves adjusting the icing detection threshold value (i.e., the green dashed line in Fig. 4.7). However, this manual threshold adjustment necessitates domain-specific expertise for parameter tuning and user intervention. Future research could explore automating hyperparameter optimization, potentially leveraging reinforcement learning techniques to dynamically adapt model settings based on environmental context and operational feedback. Furthermore, enhancing the adaptability and generalizability of the proposed model to accommodate diverse wind turbine types and varying external conditions relevant to blade icing remains a crucial objective. We defer these crucial extensions to future research efforts aimed at bolstering the robustness, adaptability, and practical utility of the DCISS framework.

## 4.5 Conclusion

In this research, we have presented a novel Deep Class-Imbalanced Semi-Supervised (DCISS) model specifically tailored for estimating blade icing conditions on wind turbines. To effectively address the inherent class imbalance challenge, we implemented a prototypical network architecture designed to re-balance class representations for both labeled and unlabeled data. The proposed prototypical network not only facilitates feature re-balancing but also enables class inference for unlabeled data by quantifying the similarity between prototypes derived from labeled and unlabeled samples within a latent feature space. To further enhance feature extraction capabilities, we integrated a channel attention module into the feature encoder component. Finally, we conducted a comprehensive empirical evaluation of the proposed DCISS model, encompassing comparative benchmarking against state-of-the-art baselines, ablation studies, and real-world online detection assessments. The experimental results unequivocally demonstrate the superior performance and practical effectiveness of the proposed DCISS model for robust and accurate blade icing detection in wind turbine systems.

## References


1. Ibrahim G, Pope K, Muzychka Y (2018) Effects of blade design on ice accretion for horizontal axis wind turbines. J Wind Eng Ind Aerodyn 173:39–52
2. Wang Y, Xu Y, Lei Y (2018) An effect assessment and prediction method of ultrasonic de-icing for composite wind turbine blades. Renew Energy 118:1015–1023
3. Wei K, Yang Y, Zuo H, Zhong D (2020) A review on ice detection technology and ice elimination technology for wind turbine. Wind Energy 23(3):433–457
4. Saleh S, Ahshan R, Moloney C (2012) Wavelet-based signal processing method for detecting ice accretion on wind turbines. IEEE Trans Sustain Energy 3(3):585–597
5. Parent O, Ilinca A (2011) Anti-icing and de-icing techniques for wind turbines: critical review. Cold Reg Sci Technol 65(1):88–96
6. Zhang M, Yu J, Chen R, Liu Q, Liu J, Song D, Liu P, Gao L, Wang J (2019) Highly transparent and robust slippery lubricant-infused porous surfaces with anti-icing and anti-fouling performances. J Alloys Compd 803:51–60
7. Peng C, Xing S, Yuan Z, Xiao J, Wang C, Zeng J (2012) Preparation and anti-icing of superhydrophobic pvdf coating on a wind turbine blade. Appl Surface Sci 259:764–768
8. Fakorede O, Feger Z, Ibrahim H, Ilinca A, Perron J, Masson C (2016) Ice protection systems for wind turbines in cold climate: characteristics, comparisons and analysis. Renew Sustain Energy Rev 65:662–675
9. Muñoz CQG, Márquez FPG, Tomás JMS (2016) Ice detection using thermal infrared radiometry on wind turbine blades. Measurement 93:157–163
10. Zeng J, Song B (2017) Research on experiment and numerical simulation of ultrasonic de-icing for wind turbine blades. Renew Energy 113:706–712
11. Gao L, Liu Y, Ma L, Hu H (2019) A hybrid strategy combining minimized leading-edge electric-heating and superhydro-/ice-phobic surface coating for wind turbine icing mitigation. Renew Energy 140:943–956
12. Yuan B, Wang C, Jiang F, Long M, Yu PS, Liu Y (2019) Waveletfcnn: a deep time series classification model for wind turbine blade icing detection. arXiv preprint arXiv:1902.05625
13. Jolin N, Bolduc D, Swytink-Binnema N, Rosso G, Godreau C (2019) Wind turbine blade ice accretion: a correlation with nacelle ice accretion. Cold Reg Sci Technol 157:235–241
14. Pryor SC, Barthelmie R (2010) Climate change impacts on wind energy: a review. Renew Sustain Energy Rev 14(1):430–437
15. Shu L, Qiu G, Hu Q, Jiang X, McClure G, Yang H (2018) Numerical and field experimental investigation of wind turbine dynamic de-icing process. J Wind Eng Ind Aerodyn 175:90–99
16. Corradini ML, Cristofaro A, Pettinari S (2016) A model-based robust icing detection and estimation scheme for wind turbines. In: 2016 European control conference (ECC). IEEE, pp 1451–1456
17. Cheng X, Li G, Ellefsen AL, Chen S, Hildre HP, Zhang H (2020) A novel densely connected convolutional neural network for sea-state estimation using ship motion data. IEEE Trans Instrum Meas 69(9):5984–5993

18. Jiménez AA, Márquez FPG, Moraleda VB, Muñoz CQG (2019) Linear and nonlinear features and machine learning for wind turbine blade ice detection and diagnosis. Renew Energy 132:1034–1048
19. Zhang L, Liu K, Wang Y, Omariba ZB (2018) Ice detection model of wind turbine blades based on random forest classifier. Energies 11(10):2548
20. Liu Y, Cheng H, Kong X, Wang Q, Cui H (2019) Intelligent wind turbine blade icing detection using supervisory control and data acquisition data and ensemble deep learning. Energy Sci Eng 7(6):2633–2645
21. Jiang G, He H, Yan J, Xie P (2018) Multiscale convolutional neural networks for fault diagnosis of wind turbine gearbox. IEEE Trans Ind Electron 66(4):3196–3207
22. Pang G, Shen C, Cao L, Hengel AVD (2021) Deep learning for anomaly detection: a review. ACM Comput Surv (CSUR) 54(2):1–38
23. Liu FT, Ting KM, Zhou Z-H (2008) Isolation forest. In: 2008 eighth IEEE international conference on data mining. IEEE, pp 413–422
24. Breunig MM, Kriegel H-P, Ng RT, Sander J (2000) Lof: identifying density-based local outliers. In: Proceedings of the 2000 ACM SIGMOD international conference on management of data, pp 93–104
25. Pang G, Cao L, Chen L, Liu H (2018) Learning representations of ultrahigh-dimensional data for random distance-based outlier detection. In: Proceedings of the 24th ACM SIGKDD international conference on knowledge discovery & data mining, pp 2041–2050
26. Pang G, Shen C, van den Hengel A (2019) Deep anomaly detection with deviation networks. In: Proceedings of the 25th ACM SIGKDD international conference on knowledge discovery & data mining, pp 353–362
27. Ruff L, Vandermeulen RA, Görnitz N, Binder A, Müller E, Müller K-R, Kloft M (2019) Deep semi-supervised anomaly detection. arXiv preprint arXiv:1906.02694
28. Orsenigo C, Vercellis C (2010) Combining discrete svm and fixed cardinality warping distances for multivariate time series classification. Pattern Recognit 43(11):3787–3794
29. Baydogan MG, Runger G, Tuv E (2013) A bag-of-features framework to classify time series. IEEE Trans Pattern Anal Mach Intell 35(11):2796–2802
30. Schäfer P (2016) Scalable time series classification. Data Mining Knowl Discov 30(5):1273–1298
31. Zheng Y, Liu Q, Chen E, Ge Y, Zhao JL (2016) Exploiting multi-channels deep convolutional neural networks for multivariate time series classification. Front Comput Sci 10(1):96–112
32. Karim F, Majumdar S, Darabi H, Harford S (2019) Multivariate lstm-fcns for time series classification. Neural Netw 116:237–245
33. Cheng X, Li G, Skulstad R, Chen S, Hildre HP, Zhang H (2019) Modeling and analysis of motion data from dynamically positioned vessels for sea state estimation. In: 2019 international conference on robotics and automation (ICRA). IEEE, pp 6644–6650
34. González M, Bergmeir C, Triguero I, Rodríguez Y, Benítez JM (2018) Self-labeling techniques for semi-supervised time series classification: an empirical study. Knowl Inf Syst 55(2):493–528
35. Gianniou P, Liu X, Heller A, Nielsen PS, Rode C (2018) Clustering-based analysis for residential district heating data. Energy Convers Manage 165:840–850
36. Chen Y, Hu B, Keogh E, Batista GE (2013) Dtw-d: time series semi-supervised learning from a single example. In: Proceedings of the 19th ACM SIGKDD international conference on Knowledge discovery and data mining, pp 383–391
37. Marussy K, Buza K (2013) Success: a new approach for semi-supervised classification of time-series. In: International conference on artificial intelligence and soft computing. Springer, pp 437–447
38. Zhang X, Gao Y, Lin J, Lu C-T (2020) Tapnet: multivariate time series classification with attentional prototypical network. In: AAAI, pp 6845–6852
39. Oliver A, Odena A, Raffel CA, Cubuk ED, Goodfellow I (2018) Realistic evaluation of deep semi-supervised learning algorithms. In: Advances in neural information processing systems, pp 3235–3246

40. Li S, Wang Z, Zhou G, Lee SYM (2011) Semi-supervised learning for imbalanced sentiment classification. In: Twenty-second international joint conference on artificial intelligence. Citeseer
41. Stanescu A, Caragea D (2014) Semi-supervised self-training approaches for imbalanced splice site datasets. In: Proceedings of the sixth international conference on bioinformatics and computational biology, BICoB 2014, pp 131–136
42. Huang C, Wu X, Zhang X, Lin S, Chawla NV (2019) Deep prototypical networks for imbalanced time series classification under data scarcity. In: Proceedings of the 28th ACM international conference on information and knowledge management, pp 2141–2144
43. Yang Y, Xu Z (2020) Rethinking the value of labels for improving class-imbalanced learning. Adv Neural Inf Process Syst 33
44. Pouyanfar S, Tao Y, Mohan A, Tian H, Kaseb AS, Gauen K, Dailey R, Aghajanzadeh S, Lu Y-H, Chen S-C (2018) Dynamic sampling in convolutional neural networks for imbalanced data classification. In: IEEE conference on multimedia information processing and retrieval (MIPR). IEEE 2018, pp 112–117
45. Buda M, Maki A, Mazurowski MA (2018) A systematic study of the class imbalance problem in convolutional neural networks. Neural Netw 106:249–259
46. Lin T-Y, Goyal P, Girshick R, He K, Dollár P (2017) Focal loss for dense object detection. In: Proceedings of the IEEE international conference on computer vision, pp 2980–2988
47. Cui Y, Jia M, Lin T-Y, Song Y, Belongie S (2019) Class-balanced loss based on effective number of samples. In: Proceedings of the IEEE conference on computer vision and pattern recognition, pp 9268–9277
48. Cao K, Wei C, Gaidon A, Arechiga N, Ma T (2019) Learning imbalanced datasets with label-distribution-aware margin loss. In: Advances in neural information processing systems, pp 1567–1578
49. He K, Zhang X, Ren S, Sun J (2016) Deep residual learning for image recognition. In: Proceedings of the IEEE conference on computer vision and pattern recognition, pp 770–778
50. Snell J, Swersky K, Zemel R (2017) Prototypical networks for few-shot learning. In: Advances in neural information processing systems, pp 4077–4087
51. Hayat M, Khan S, Zamir W, Shen J, Shao L (2019) Max-margin class imbalanced learning with gaussian affinity. In: ICCV
52. Wang Z, Yan W, Oates T (2017) Time series classification from scratch with deep neural networks: A strong baseline. In: 2017 international joint conference on neural networks (IJCNN). IEEE, pp 1578–1585
53. Hu J, Shen L, Sun G (2018) Squeeze-and-excitation networks. In: Proceedings of the IEEE conference on computer vision and pattern recognition, pp 7132–7141

# Part II
# Graph Convolutional Neural Network-Based Methods

As highlighted in the preceding part introduction, Part II pivots to explore the application of Graph Neural Network (GNN)-based methods for blade icing detection. This shift recognizes a crucial aspect of wind turbine sensor data: its inherent interconnected nature. While Part I effectively leveraged temporal patterns with TCNs, Part II delves into the potential of GNN architectures to explicitly model the complex relationships between sensors, further enhancing icing detection accuracy. These innovative methods harness the power of graph neural networks to capture spatial-temporal dependencies and significantly improve feature learning directly from graph-structured sensor data. By leveraging graph convolutions and attention mechanisms, GNNs demonstrate a remarkable ability to understand the intricate interplay between sensor readings across the wind turbine system, leading to more nuanced and robust icing detection. Each method presented in Part II addresses a unique challenge in effectively leveraging graph neural networks for blade icing detection, and is purposefully designed to contribute significantly to enhancing model performance and overall robustness in real-world scenarios.

The core objective of Part II is to provide a comprehensive understanding of how graph neural network methodologies can be effectively employed for advanced blade icing detection. Moving beyond purely temporal analysis, these GNN-based methods are specifically engineered to model the complex relationships between sensors, effectively leveraging the spatial-temporal information intrinsically present in wind turbine sensor data. By employing advanced techniques such as graph convolutions and attention mechanisms within the GNN framework, the models are able to capture intricate patterns and dependencies that arise from sensor interconnections, leading to a significant improvement in the accuracy and robustness of icing detection. Furthermore, Part II introduces novel approaches to address practical challenges inherent in graph-based modeling, such as data imbalance and computational complexity, ensuring the real-world applicability of these advanced techniques. For example, Chap. 5 introduces the Attention-Enhanced Graph Convolutional Neural Network (GTAN), a novel GNN architecture designed to explicitly model sensor relationships. GTAN utilizes graph convolutional networks (GCNs) to capture spatial relationships and integrates temporal attention mechanisms to capture temporal dynamics, adaptively analyzing sensor correlations and prioritizing relevant temporal segments within the graph structure. Expanding on feature richness, Chap. 6 proposes a Wavelet-Empowered Graph Convolutional Neural Network (MWGCN). This innovative GNN method combines multiscale wavelet decomposition with graph neural networks, leveraging wavelet analysis for feature extraction across different scales and Temporal Graph Convolutional Networks (TGCNs) for modeling intervariable correlations within the graph, further enhanced by scale-attention to capture inter-scale dependencies. Addressing practical data realities, Chap. 7 presents a Class-Imbalanced Graph-Temporal Convolutional Neural Network. This specialized GNN architecture builds upon graph convolutional networks and incorporates spatial-temporal feature learning, while crucially integrating a distance-based classifier specifically designed to mitigate bias towards the majority class in imbalanced datasets, a common issue in real-world sensor data.

Each chapter included in Part II meticulously details a unique Graph Neural Network approach, substantiated by rigorous experimental evaluations conducted on publicly available benchmark datasets, as well as real-world Supervisory Control and Data Acquisition (SCADA) data collected from operational wind turbine systems. The experimental findings comprehensively presented throughout this part emphatically underscore the considerable potential and practical applicability of Graph Neural Networks within the context of blade icing detection. This lays a robust foundation for the realization of more advanced and effective icing detection methodologies, moving towards a more holistic understanding of wind turbine systems. This research significantly enriches the wider understanding of graph-based modeling strategies tailored for deployment in complex and interconnected systems like wind turbines operating in challenging environments.

# Chapter 5
# Attention-Enhanced Graph Convolutional Neural Network for Blade Icing Detection

## 5.1 Introduction

The significance of electrical energy is continually escalating, underscored by an approximate annual demand surge of 3.6% over the past three decades [1]. Renewable energy sources are indispensable for meeting this demand in a sustainable manner, with wind energy emerging as a pivotal contributor, notably devoid of greenhouse gas emissions. However, the inherent fluctuations in wind velocity and the geographical constraints of regions with consistently robust winds curtail its potential as a universally applicable energy solution [2]. Notwithstanding these limitations, wind turbines are being increasingly deployed globally as a crucial element of the worldwide energy portfolio [3–5]. A primary operational challenge, especially for wind farms situated in frigid environments such as mountaintops, is blade icing, resulting from the atmospheric freezing of moisture [6]. It is estimated that approximately one-third of wind turbines operating in cold climates across the globe are impacted by blade icing. This phenomenon instigates a cascade of detrimental consequences: diminished aerodynamic efficiency leading to reduced power output, heightened structural stress potentially causing equipment failure, and critically, safety risks that can result in injuries or fatalities [7–10]. Consequently, the prompt and dependable detection of blade icing is not merely an operational refinement but a fundamental imperative for ensuring the secure, effective, and uninterrupted power generation of wind turbines in colder regions.

To alleviate the adverse impacts of blade icing, the deployment of icing sensors has become standard practice. These sensors are implemented to ascertain the presence of ice on wind turbine blade surfaces, and subsequently, to activate and validate the effectiveness of blade de-icing systems [11]. Homola, M. C. et al. presented a thorough overview of this field, cataloging 24 direct and 5 indirect methods for icing detection based on sensors [11]. While offering diverse methodologies, sensor-based approaches encounter several limitations. For example, temperature-based sensors, while seemingly straightforward, often struggle to detect thin ice layers due to the



X. Cheng et al., *Computational Methods for Blade Icing Detection of Wind Turbines*, Engineering Applications of Computational Methods 24,
https://doi.org/10.1007/978-981-96-6763-5_5

subtle temperature differentials, leading to potential false negatives [12]. Similarly, techniques that rely on reflected light for icing detection, such as visual inspection systems or optical sensors, are primarily effective in identifying frost or opaque ice formations on blade surfaces. They often find it challenging to detect transparent thin ice layers, which, despite their transparency, can still significantly compromise aerodynamic performance without substantially altering light reflection [13]. Zhixin, F. U. et al. sought to enhance detection reliability by developing a wireless sensor network-based icing detection method for wind turbines [14]. This approach utilizes distributed sensors to monitor both blade surface temperature and humidity, aiming to improve icing prediction accuracy by considering environmental variables. While this method demonstrated improved reliability under certain conditions, wireless sensor networks can be vulnerable to communication disruptions in harsh weather conditions, and the accuracy remains dependent on the sensors' capacity to detect subtle changes indicative of early icing. Kim D. et al. explored the application of fiber-optic sensors for icing detection, monitoring both temperature and blade surface deformation [15]. Fiber-optic sensors offer advantages in sensitivity and resistance to electromagnetic interference, potentially enabling rapid and precise icing detection via minute temperature or strain variations. However, traditional physical methods, in general, often face constraints related to cost-effectiveness and overall precision. The installation and upkeep of extensive sensor networks can be costly, particularly in remote and harsh wind farm locations. Furthermore, the inherent accuracy limitations of individual sensor types, their susceptibility to environmental noise, and the difficulties in ensuring complete blade coverage with sensors remain significant concerns. Moreover, numerous active sensor-based de-icing systems consume extra energy and may potentially affect the long-term mechanical integrity of wind turbine blades due to repeated thermal or mechanical stresses. These limitations of traditional physical methods, notably the cost and accuracy trade-offs, and the potential for invasive installation and maintenance, have spurred the exploration of alternative, data-driven methodologies.

In recent years, with the increasing sophistication and widespread adoption of Supervisory Control and Data Acquisition (SCADA) systems in wind farms, a growing research interest has emerged in leveraging data-driven approaches for wind turbine blade icing detection. Machine learning (ML) and deep learning (DL) techniques have gained prominence as promising data-driven alternatives to conventional sensor-based methods. These approaches present the potential for automating the icing detection process by learning directly from historical operational data captured by SCADA systems, thereby enhancing wind farm safety and dependability while simultaneously reducing reliance on expensive and maintenance-intensive external sensors. Furthermore, data-driven methods can potentially offer more refined and predictive icing detection capabilities by capturing intricate patterns and interdependencies within the vast quantities of operational data. However, the application of machine learning and deep learning to wind turbine blade icing detection also introduces considerable challenges. A primary challenge is the inherent imbalance within the datasets. Icing events occur relatively infrequently compared to normal, non-icing operating conditions, resulting in highly skewed datasets where negative samples (non-icing) significantly outnumber positive samples (icing). This imbalance can severely bias standard machine learning and deep learning algorithms, which tend to be optimized for balanced datasets, leading to suboptimal performance in detecting the minority class (icing events).

Within the domain of machine learning for icing prediction, Tao et al. introduced the Stacked-XGBoost model, a sophisticated ensemble method that ingeniously combines fundamental ice physics principles with hybrid features derived from both meteorological and SCADA datasets [16]. By incorporating physics-informed features, their model aimed to enhance the generalizability of icing detection across diverse wind turbine locations and operational scenarios. While XGBoost models are recognized for their high predictive capacity and ability to manage complex non-linear relationships, they are also computationally demanding and necessitate meticulous feature engineering. The process of manually crafting effective features still requires significant domain expertise and can be time-consuming. Moreover, complex models like XGBoost can be susceptible to overfitting, especially when trained on limited or noisy datasets, potentially hindering their performance on unseen data. Guo et al. also employed an XGBoost-based model, concentrating on readily available SCADA data such as output power and rotor speed for icing detection [17]. Their work distinguished itself by incorporating a Sequential Probability Ratio Test (SPRT) for residual analysis. This innovative integration of SPRT aimed to provide early detection of incipient icing events by statistically analyzing deviations from anticipated operational behavior, offering a more proactive warning system compared to purely threshold-based or sensor-triggered methods. However, despite these advancements, XGBoost models, even with SPRT enhancement, can still be challenging to interpret, acting somewhat as "black boxes", which can limit the understanding of the underlying icing mechanisms and potentially impede model refinement and optimization. Furthermore, the performance of XGBoost models remains heavily influenced by the quality and relevance of the input features, and their ability to detect subtle, early-stage icing anomalies, particularly in noisy real-world SCADA data, remains a topic of ongoing research. Mucchielli, P., et al. explored a cost-sensitive, real-time downtime detection technique for wind turbines, leveraging eigen perturbation approaches in conjunction with supervised machine learning [18]. Their methodology demonstrated precise detection and classification of various downtime events, showcasing the potential of machine learning for broader wind turbine condition monitoring beyond solely icing. However, while effective for downtime detection, the direct applicability and performance of eigen perturbation-based methods specifically for the nuanced detection of blade icing, with its often subtle and gradual onset, necessitates further investigation.

Expanding beyond traditional machine learning, deep learning methods have also been explored for wind turbine icing detection. Hacıefendioğlu, Kemal, et al. proposed a unique approach employing image processing and deep learning paradigms for wind turbine icing detection [19]. This technique utilized hyperspectral imaging data, combining advanced image processing with semantic segmentation techniques to directly discern and quantify icing on blade surfaces from visual data. While demonstrating notable success in detecting turbine icing through visual means, image-based methods often encounter limitations in practical deployment. The acquisition of high-quality hyperspectral images can be costly and require specialized equipment, and the performance can be affected by weather conditions, visibility, and the need for continuous image acquisition and processing. Furthermore, the scalability and real-time applicability of image-based methods for large-scale wind farm monitoring, especially compared to readily available SCADA data, present logistical and computational challenges. In the context of addressing data scarcity and privacy concerns in distributed wind farms, Cheng et al. introduced a heterogeneous federated learning (FL) framework for blade icing detection [20]. Federated learning enables collaborative model training across multiple wind turbines or wind farms without directly sharing sensitive raw data, offering a privacy-preserving approach. However, federated learning introduces its own set of challenges, including dealing with heterogeneous data distributions across different turbines, managing communication overhead between distributed agents, and ensuring the robustness and security of the federated learning process itself. Expanding the scope of deep learning architectures, Zhao et al. developed the MTAD-GAT model [21], employing a graph-structured network with an attention mechanism for anomaly detection in multivariate time-series data. This model adeptly identifies anomalies in both temporal and feature dimensions, representing a significant advancement in anomaly detection technology applicable to complex systems like wind turbines. While MTAD-GAT demonstrates promising capabilities for general anomaly detection, its direct application and optimization for the specific task of wind turbine blade icing detection, particularly in the context of imbalanced data and the need for high precision in identifying icing events, warrants further investigation.

In summary, the literature review emphasizes the significant potential of both machine learning and deep learning methodologies in enhancing wind turbine blade icing detection, with a consistent research emphasis on achieving high accuracy and robustness. However, a critical gap remains inadequately addressed across these data-driven approaches: the effective management of imbalanced data, a pervasive characteristic of real-world icing detection scenarios. The infrequent occurrence of icing events, in contrast to normal operation, leads to datasets heavily skewed towards non-icing conditions. This data imbalance poses a substantial challenge for most standard machine learning and deep learning algorithms, which can become biased towards the majority class and exhibit poor performance in detecting the minority class of critical icing events. Addressing this imbalanced data challenge is therefore paramount for developing practical and reliable data-driven icing detection systems. Furthermore, the high dimensionality of SCADA data, often containing redundant, irrelevant, and noisy information, can further complicate the task of effective feature extraction and model training, potentially leading to overfitting and reduced generalization performance. The computational cost associated with processing large volumes of high-dimensional SCADA data and training complex deep learning models also necessitates efficient model architectures and optimization strategies.

To surmount the limitations of existing methods and specifically address the challenges associated with the detection of wind turbine blade icing in imbalanced datasets, this paper introduces the Graph Temporal Attention Network (GTAN).

Central to our approach is a specifically designed imbalanced classifier, intended to mitigate the detrimental effects of skewed data distribution and enhance the model's capacity to accurately detect the minority class of icing events. Furthermore, to effectively process time-series SCADA data and capture temporal dependencies crucial for icing detection, a Temporal Attention (TA) module is integrated into the feature extractor. This TA module enhances the model's ability to differentiate between instances of blade icing and non-icing conditions by selectively attending to the most pertinent temporal segments of the input data. For wind farm operators, the GTAN model offers a promising avenue to optimize maintenance strategies and enhance operational efficiency. By providing more accurate and timely icing detection, GTAN can potentially lead to substantial cost reductions by enabling proactive and targeted maintenance interventions, streamlining maintenance schedules, and allocating resources precisely where and when they are most needed. This strategic approach holds the potential to significantly lessen the economic impact of blade icing incidents, aligning with the operational efficiency and profitability objectives of modern wind farms.

The contributions of this research can be summarized as follows:

1. To effectively manage time-series SCADA data and address the critical limitations of current methods in imbalanced icing detection scenarios, we incorporate the Temporal Attention (TA) mechanism into the Graph Attention Network (GAT) model. This results in the novel Temporal-GAT model, enabling GTAN to learn discriminative features and achieve precise blade icing detection, particularly in challenging imbalanced datasets. Furthermore, a dedicated Feature Extractor module is meticulously designed to amplify the inherent differences between ice and non-ice data representations, thereby enhancing the model's competitive performance and robustness, especially under conditions of high data imbalance.
2. The proposed GTAN model is rigorously evaluated using real-world SCADA data collected from three operational wind turbines. To comprehensively assess the model's performance under varying data distributions, we generate three distinct datasets from the raw sensor data, each characterized by different degrees of class balance. The superiority and practical significance of the GTAN model are demonstrably validated through extensive comparative analysis against a range of established baseline networks using the SCADA datasets. Additionally, to further ascertain the effectiveness of the designed modules and the overall robustness of the model, comprehensive sensitivity analysis and robustness analysis are meticulously conducted, providing deeper insights into the model's behavior and reliability.

The subsequent sections of this paper are organized as follows: Section II provides a comprehensive review of relevant literature, delving deeper into data-driven wind turbine blade icing detection methodologies. Section III presents a detailed explanation of the proposed GTAN model, including a thorough description of the Feature Extractor module and the core Temporal-GAT model architecture. Section IV outlines the experimental methodology and presents a comprehensive evaluation of the proposed method using the three SCADA datasets, including comparative results, sensitivity analysis, and robustness analysis. Finally, Section V concludes the paper by summarizing the key contributions and significant findings of this study, highlighting its implications for wind energy operations and future research directions.

## 5.2 Methodology

This section details the GTAN model, emphasizing its core components and key fusion techniques. Specifically, we will discuss data preprocessing methods employed to ensure data continuity and quality for subsequent model training and evaluation. Additionally, we will explore the capabilities of the GAT model, highlighting its proficiency in extracting vital information from both temporal and sensor feature dimensions, thereby enhancing its overall efficacy.

### *5.2.1 Overview*

To address the challenge of imbalanced data in wind turbine blade icing detection, our methodology incorporates both data-level and model-level strategies. We employ preprocessing techniques to generate datasets exhibiting varying degrees of imbalance. At the model level, we emphasize under-represented samples using the Focalloss function and the Nearest Class Mean (NCM) classifier. As depicted in Fig. 5.1, the preprocessed data undergoes enhancement via a U-shaped Feature Extractor employing one-dimensional convolutional layers. This enhanced data is then fed into the GTAN model to effectively capture temporal relationships. Ultimately, the NCM classifier, utilizing GTAN's extracted features, assesses icing probabilities on wind turbine blades. This methodological framework offers a comprehensive solution to this significant challenge in real-world applications.

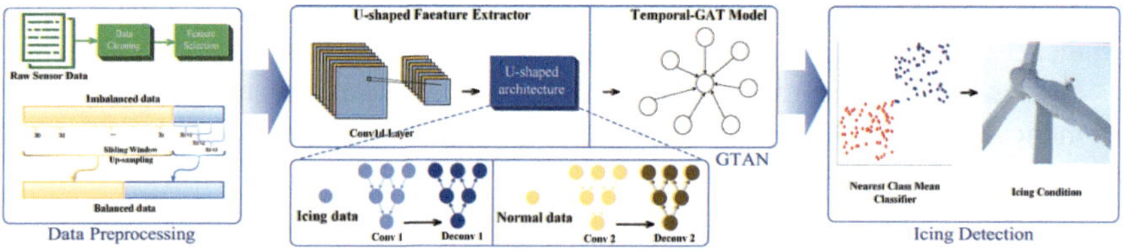


**Fig. 5.1** Framework for wind turbine blade icing detection. The photograph of iced wind turbine blades is sourced from [22]

### 5.2.2 Data Preprocessing

Data preprocessing is a crucial phase in model training, encompassing meticulous data cleansing, astute feature selection, and judicious management of data imbalance. In this study, the initial focus was on purifying the raw sensor data to mitigate the impact of anomalies and noise. Drawing upon domain expertise and utilizing Pearson correlation analysis (PCA), we identified 26 key features relevant to wind turbine blade icing, thereby reducing data redundancy. Conventional re-balancing techniques, such as undersampling the majority class or oversampling the minority class, are common; however, these methods can respectively lead to the loss of valuable data or induce model overfitting [23]. An alternative approach involves extracting features from the existing dataset and generating synthetic instances of the same class, as exemplified by DeepSMOTE [24]. However, practical application in this project revealed that the data synthesized by DeepSMOTE was consistently of lower quality than the actual sensor data, resulting in suboptimal training outcomes.

To resolve these issues, we implemented the Sliding Window Upsampling (SWU) technique. This method is both straightforward and effective, preserving the temporal sequence of the data. Unlike other methods, SWU employs overlapping windows for the minority icing data and non-overlapping windows for the majority normal data, effectively increasing the sample size of the minority class, as shown in Fig. 5.2. This process generated $t$ independent samples from normal data and $n$ augmented samples from icing data, with $n$ depending on the window's overlap extent or stride, and potentially equaling or exceeding $t$ when necessary. It is important to note that the SWU process is applied exclusively to segments of the same class within continuous time intervals. For example, if icing data is present in separate intervals, SWU is applied within each interval, thus maintaining temporal coherence. This preprocessing strategy has yielded a dataset that is not only of high quality but also specifically tailored for time-series analysis, providing a robust foundation for subsequent model training and performance evaluation.

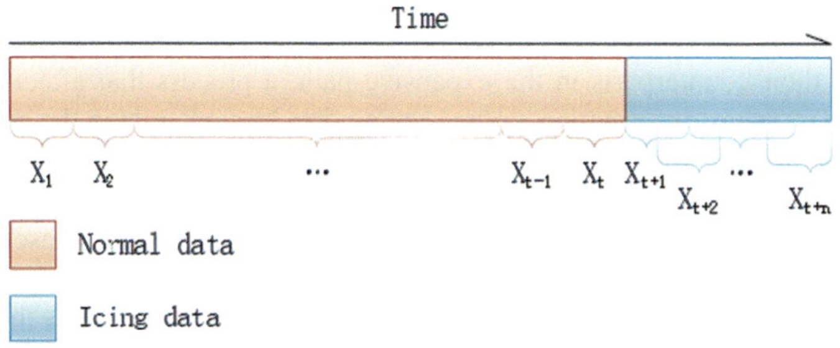


**Fig. 5.2** Illustrative example of SWU

### 5.2.3 *Graph Temporal Attention Network (GTAN)*

In this section, we provide a detailed description of the GTAN model, a central component of our methodology. The GTAN model consists of two primary modules: a U-shaped Feature Extractor and a Temporal-GAT model.

**(1) U-shaped Feature Extractor**

Given the variable environmental conditions under which wind turbines typically operate and the inherent advantages of CNNs in automatic feature learning for data comprehension, we utilized a one-dimensional convolution layer to extract advanced features at each time step of the data. Each data feature serves as a channel input to the convolution layer.

The Convolution layer comprises a sequential stack of operations, including padding, basic convolution, and Rectified Linear Unit (ReLU), defined as follows:

$$\begin{aligned} \mathbf{Y} &= \mathrm{PADDING}(\mathbf{X}), \\ \mathbf{Y} &= \mathrm{CONV}(\mathbf{Y}, \mathbf{W}, \mathbf{b}), \\ \mathbf{Y} &= \mathrm{ReLU}(\mathbf{Y}), \end{aligned} \tag{5.1}$$

where $\mathbf{X} \in \mathbb{R}^{\mathbf{T}\times\mathbf{M}}$ represents the weighted input, $\mathbf{Y}$ denotes the intermediate features, and $\mathbf{W}$ and $\mathbf{b}$ are the trainable parameters within the 1D CNN.

The adoption of the U-shaped architecture in deep learning models is becoming increasingly significant for pixel-level tasks, including image segmentation and generation [25, 26]. This architecture is characterized by its tightly integrated contracting and expansive paths [27]. The contracting path employs a series of convolutional and pooling layers to progressively downsample the input image, effectively extracting and compressing key feature information. Conversely, the expansive path reverses this process, gradually restoring the image to its original dimensions. This distinctive structural design significantly enhances the model's ability to assimilate deep contextual information, derived from the network's deeper layers, as well as fine-grained details, originating from the shallower layers. A defining characteristic of the U-shaped architecture is the fusion of outputs from each level of the contracting path with their counterparts in the expansive path, a process that effectively preserves and transmits information, counteracting data loss inherent in the contraction phase. Additionally, the implementation of data augmentation strategies within the U-shaped architecture substantially improves its effectiveness on smaller training datasets, a crucial element in fields where extensively annotated datasets are not readily available, such as in wind turbine blade icing detection.

In our research, we have adopted the U-shaped architecture to enhance the model's capability to differentiate between positive and negative samples. The Conv module within this framework is designed for abstracting features from data patches, while the Deconv module is responsible for image reconstruction (interpreting the feature matrix as an image), thus highlighting the disparities between positive and negative samples. The contracting and expansive pathways of the U-shaped architecture are analogous to the Conv and Deconv modules, respectively. Leveraging the strengths of the U-shaped architecture enables the Conv module to abstract features that are subsequently utilized by the Deconv module for sample reconstruction, thereby facilitating a more pronounced differentiation between positive and negative samples.

Within the Conv module, convolutional layers use filters to identify correlations within the input data. These filters traverse the input data, performing element-wise multiplications and aggregating the results to produce a feature map. As the data progresses through multiple convolutional layers, the Conv module progressively acquires more complex and abstract representations of the input data.

The output of the Conv module $\mathbf{a}_{\text{out}}$ is defined as:

$$\mathbf{a}_{\text{out}} = f(\mathbf{W} \times \mathbf{a}_{\text{in}} + \mathbf{b}), \tag{5.2}$$

where $f$ represents the activation function, $\times$ denotes the convolution operation, and $\mathbf{W}$ and $\mathbf{b}$ are the parameters learned in the Conv module.

As a symmetric operation to the Conv module, we employed the following form of the deconvolution operation:

$$\mathbf{a}_{\text{out}} = f(\mathbf{W} \otimes \mathbf{a}_{\text{in}} + \mathbf{b}), \tag{5.3}$$

where $\otimes$ represents the deconvolution operation.

**(2) Temporal-GAT**

The integration of the TA module within the Temporal-GAT model significantly enhances the system's effectiveness in distinguishing icing from non-icing conditions on wind turbine blades [22]. This integration is a strategic refinement of the TA module's capabilities within the GAT framework, representing an innovative approach in temporal data analysis. Designed to evaluate the significance of each sensor and time step, the TA module equips the Temporal-GAT model to prioritize essential information for accurate icing condition identification. This targeted approach, deviating from the conventional practice of uniform sensor data consideration, substantially improves the model's analytical precision.

Strategically positioned before the GAT layer, the TA module facilitates the autonomous identification and extraction of distinctive features from unprocessed sensor data, surpassing traditional methodologies. This advancement enables the Temporal-GAT model to uniquely emphasize characteristics that are strongly indicative of icing conditions, a crucial attribute given the intricate nature of icing events and their dependence on specific environmental and operational parameters.

The TA module's effectiveness within the Temporal-GAT model arises from its capacity to enhance data-driven models that rely on input data quality. By processing and weighting sensor data, the TA module is instrumental in isolating critical sensors and time steps, channeling this refined data into the GAT layer for sophisticated feature extraction and analysis. This methodology ensures that the Temporal-GAT model receives relevant, high-quality data while maintaining a strong emphasis on temporal dynamics and sensor-specific nuances, thereby markedly improving the model's accuracy and reliability in detecting blade icing conditions.

Considering the decay of temporal correlations over extended periods, our methodology includes pruning edges associated with long time intervals, retaining only connections to the nearest 20% of nodes in both preceding and succeeding time frames. This edge pruning strategy, inspired by the TA module's focus on pertinent sensors and time steps, ensures that the GAT layer concentrates on the most influential temporal relationships. This focus is critical, given the significant impact of both current and historical environmental and operational data on wind turbine blade icing conditions.

The specific implementation of the GAT layer is detailed as follows:

Consider a graph with $n$ nodes, represented as $\mathbf{v}_1, \mathbf{v}_2, \mathbf{v}_3, \ldots, \mathbf{v}_n$, where $\mathbf{v}_i$ is the feature vector of each node. The GAT layer computes the output representation of each node using the following formula:

$$\begin{aligned} \mathbf{h}_i &= \sigma\Big(\sum_{j \in \mathbf{N}(\mathbf{v}_i)} \alpha_{ij}\mathbf{v}_j\Big), \\ \mathbf{e}_{ij} &= \text{LeakyReLu}(\mathbf{w}^T \times (\mathbf{v}_i \oplus \mathbf{v}_j)), \\ \alpha_{ij} &= \frac{exp(\mathbf{e}_{ij})}{\sum_{l \in \mathbf{N}(\mathbf{v}_i)} exp(\mathbf{e}_{il})}, \end{aligned} \tag{5.4}$$

where $\mathbf{h}_i$ represents the output representation of node $i$, $\sigma$ denotes the sigmoid activation function; $\alpha_{ij}$ is the attention score between node $i$ and node $j$, $\mathbf{N}(\mathbf{v}_i)$ is the set of neighbors of node $i$, $\mathbf{w}$ represents the parameters to be learned, and $\oplus$ denotes the concatenation of two nodes.

In essence, the incorporation of the TA module into the Temporal-GAT framework represents a synergistic combination, integrating the strengths of attention-driven feature discernment and graph-based temporal analysis. This integration provides a robust and sophisticated framework for precisely modeling the complex interdependencies characteristic of wind turbine blade icing detection scenarios.

### 5.2.4 Icing Detection of Wind Turbine Blades

In this study, wind turbine blade icing detection is framed as a binary classification task. However, conventional softmax classifiers exhibit limitations in addressing class-imbalanced learning problems, as highlighted by Kim et al. [28]. This is because these classifiers treat all categories equally during training, hindering the effective learning of features from minority classes. Moreover, relying on probability distribution, softmax classifiers are susceptible to the impact of class imbalance in the dataset, leading to a decision boundary biased towards majority classes. The formula for the softmax classifier is as follows:

$$\mathbf{P}(y = j|x) = \frac{e^{z_j}}{\sum_{k=1}^{K} e^{z_k}}, \tag{5.5}$$

where $z_j$ is the score for class $j$ corresponding to input $x$, $K$ is the total number of classes, and $\mathbf{P}(y = j|x)$ is the probability that the classifier assigns input $x$ to class $j$.

To overcome these limitations, we integrate the NCM classifier, which is instrumental in adjusting the decision boundary to enhance classification performance [29]. The NCM classifier, through its distance-based approach, is particularly adept at effectively capturing the characteristics of minority classes, thus mitigating the dominant influence of majority classes in the dataset. This approach is crucial in scenarios where class imbalance is prominent, such as in wind turbine blade icing detection, where non-icing instances might significantly outnumber icing events.

The operational mechanism of the NCM classifier is based on its distance-based algorithm, which assigns data samples to the class with the nearest mean value, as represented by:

$$\mathbf{y}^{\star} = \text{argmin}_y d(x, \mu_y), \quad \mathbf{y} \in \{1, 2, ..., K\}, \tag{5.6}$$

where $\mu_y$ represents the mean of class $y$, defined as $\mu_y = \frac{1}{N_y} \sum_{i:\mathbf{y}_i = y} \mathbf{x}_i$, and $N_y$ is the number of samples in class $y$. Here, $d(x, \mu_y)$ denotes the distance between a sample $x$ and the class mean $\mu_y$, and $K$ represents the total number of classes. The effectiveness of the NCM classifier is significantly dependent on the chosen distance metric, which is critical in accurately delineating class boundaries, especially in cases where the training data is skewed towards certain classes.

Furthermore, to specifically address the issue of class imbalance in our dataset, we have refined our loss function. Employing the back-propagation algorithm in conjunction with the gradient descent optimization technique, we have integrated the Focalloss function, denoted as $FL(p_t)$, into our model [30]. This function is formulated as:

$$FL(p_t) = -\alpha_t (1 - p_t)^{\gamma} \log(p_t), \tag{5.7}$$

where $p_t$ is the model's estimated probability for the ground-truth class, $\gamma$ is a focusing parameter that modulates the contribution of each sample to the loss based on the ease of classification, and $\alpha_t$ is a balancing factor for adjusting the contribution of positive and negative classes. The parameter $\gamma$ effectively reduces the loss contribution from easy examples, thereby focusing the training more on hard negatives. This approach ensures that the model pays more attention to less frequent but more informative samples, thus mitigating the issues caused by class imbalance.

For the optimization of the loss function and to facilitate efficient computation while reducing memory usage, the Adam algorithm was utilized. This algorithm is recognized for its efficiency in handling sparse gradients and adaptive learning rate modulation, making it an optimal choice for training models, particularly under class imbalance conditions.

## 5.3 Experiment

All experiments were conducted on a workstation configured with an Intel(R) Core(TM) i7-9700KF CPU, 16 GB of RAM, and an NVIDIA GeForce RTX 2070 SUPER graphics card. PyTorch was employed for the implementation of all models.

### 5.3.1 Dataset and Evaluation Metrics

In this study, wind turbine blade icing data was sourced from Goldwind Science & Technology Co., Ltd., a prominent global wind turbine manufacturer. The data was collected from operational logs of the SCADA systems of three wind turbines located in Inner Mongolia, China. The operational durations recorded for these turbines were 305.77, 695.59, and 329.28 h, respectively. Notably, the first two turbines experienced operational breaks due to scheduled shutdowns, while only the third turbine maintained continuous data logging. Consequently, data from the two turbines with intermittent sensor data were aggregated for offline training and testing purposes.

The raw sensor data was multidimensional. To refine data processing, a collaborative effort involving domain experts and PCA was undertaken, leading to the selection of 26 parameters directly relevant to blade icing, as detailed in Table 5.1. Furthermore, engineers played a crucial role in annotating instances of icing within the dataset.

To address the significant class imbalance between positive and negative samples, initially at a ratio of approximately 1:50, an imbalanced data processing methodology was applied, as detailed in Section III-B. This method resulted in the creation of three distinct datasets, designated A, B, and C, achieved by adjusting the stride of the sliding window. Consequently, the positive-to-negative sample ratios in these datasets were established at $\{1, 0.1, 0.05\}$, respectively. Following data cleaning and preprocessing procedures, the consolidated dataset comprised 1259 h of data. An 80% portion of this dataset was allocated for training, with the remaining 20% reserved for testing.

To evaluate the model's performance, we utilized the following metrics: precision, recall, accuracy, F1-score, and Matthews correlation coefficient (MCC). These metrics are defined as follows:

$$
\begin{aligned}
&\text{Precision} = \frac{\text{TP}}{\text{TP} + \text{FP}}, \quad \text{Recall} = \frac{\text{TP}}{\text{TP} + \text{FN}}, \\
&\text{Accuracy} = \frac{\text{TP} + \text{TN}}{\text{TP} + \text{FP} + \text{TN} + \text{FN}}, \\
&\text{F1} = \frac{2 \times \text{Precision} \times \text{Recall}}{\text{Precision} + \text{Recall}}, \\
&\text{MCC} = \frac{\text{TP} \times \text{TN} - \text{FP} \times \text{FN}}{\sqrt{(\text{TP} + \text{FP})(\text{TP} + \text{FN})(\text{TP} + \text{FP})(\text{TN} + \text{FN})}},
\end{aligned}
\tag{5.8}
$$

**Table 5.1** Wind turbine SCADA data parameters

| Index | Monitored variable | Index | Monitored variable |
|---|---|---|---|
| 1 | Wind velocity | 14 | Pitch motor 1 temperature |
| 2 | Generator rotational speed | 15 | Pitch motor 2 temperature |
| 3 | Active electrical power output | 16 | Pitch motor 3 temperature |
| 4 | Wind direction (compass heading) | 17 | Horizontal acceleration (Turbine structure) |
| 5 | 25-s averaged wind direction angle | 18 | Vertical acceleration (Turbine structure) |
| 6 | Yaw system position | 19 | Ambient temperature |
| 7 | Yawing rate | 20 | Nacelle internal temperature |
| 8 | Pitch angle—Blade 1 | 21 | Pitch 1 switching element temperature |
| 9 | Pitch angle—Blade 2 | 22 | Pitch 2 switching element temperature |
| 10 | Pitch angle—Blade 3 | 23 | Pitch 3 switching element temperature |
| 11 | Pitch actuator 1 speed | 24 | Pitch 1 charger DC power consumption |
| 12 | Pitch actuator 2 speed | 25 | Pitch 2 charger DC power consumption |
| 13 | Pitch actuator 3 speed | 26 | Pitch 3 charger DC power consumption |

where TP, FP, TN, and FN represent true positive, false positive, true negative, and false negative counts, respectively.

### *5.3.2 Baseline Comparison*

This section presents a comparison of the proposed wind turbine vortex ice detection method against five baseline networks to validate its performance. The MTAD_GAT model, recognized as a high-performing graph neural network, along with four conventional networks, are used for comparison. The five baseline networks and their specifications are detailed below:

1. **Long Short-Term Memory (LSTM)**: LSTM is a robust tool for modeling time-series data. In this experiment, we utilized a single-layer recurrent LSTM model for time-series modeling. The number of hidden units was chosen from $\{16, 32, 64, 128\}$, and the model exhibiting the best performance was selected for comparison.

2. **Gated Recurrent Unit (GRU)**: GRU, considered a computationally lighter alternative to LSTM with comparable performance, is examined in this experiment. A single-layer recurrent GRU model was configured, with the number of hidden units selected from {16, 32, 64, 128}, and the model demonstrating the best performance was chosen for comparison.
3. **Convolutional Neural Network (CNN)**: In this experiment, we employed a one-dimensional convolutional CNN layer. The size of the convolution kernel was selected from {3, 5, 7}, and the model achieving the best performance was chosen for comparison.
4. **Fully Convolutional Network (FCN)**: FCN has shown competitive performance in time-series classification [31]. In this experiment, we employed three convolutional layers with filter sizes of {128, 256, 128}, and the model with the best performance was selected for comparison.
5. **MTAD-GAT**: This model is an enhanced version of GAT, specifically designed for multivariate time-series anomaly detection, with a focus on icing detection on turbine blades. We experimented with various neighbor ranges for MTAD-GAT, selected from $\{\pm 0.2, \pm 0.5, \pm 1\}$, and the model with the best performance was chosen for comparison.

We standardized each baseline network in our comparison by using the same loss function and classifier as GTAN, with a consistent window size (T) set to 32 (equivalent to approximately 4 min of sensor data). To account for variations among the models in our comparison, we conducted hyperparameter tuning during the training phase for each model. This tuning covered a range of learning rates from $1e^{-5}$ to $1e^{-1}$ and an epoch range from 5 to 500. All results reported for each model are derived from the optimal hyperparameter configurations, and these results are the averages obtained from ten consecutive experiments.

Our proposed model demonstrated superior performance across different balanced datasets, as shown in Table 5.2. For Dataset A, our model improved the MCC and F1-score by 6.1 and 4.2%, respectively, compared to the best results achieved by MTAD-GAT. In comparison to the FCN model, which performed the least effectively, our model showed a substantial improvement of 29 and 13.5% in MCC and F1-score, respectively.

**Table 5.2** Comparative results on dataset A

| Models | MCC | ACC | F1 | Precision | Recall |
|---|---|---|---|---|---|
| GTAN | **59.3** | **79.4** | **78.8** | **80.8** | **79.4** |
| MTAD_GAT | 53.2 | 75.9 | 74.6 | 78.7 | 75.9 |
| FCN | 30.3 | 65.8 | 65.3 | 65.7 | 65.8 |
| CNN | 45.1 | 72.9 | 72.5 | 73.1 | 72.9 |
| LSTM | 50.5 | 74.2 | 72.5 | 78.1 | 74.2 |
| GRU | 50.5 | 74.6 | 73.2 | 77.5 | 74.6 |

As indicated in Table 5.3, although all metrics decreased in Dataset B, our model still outperformed MTAD_GAT by nearly 7.1 and 3.7% in terms of MCC and F1-score, respectively. Compared to the FCN model with the lowest performance, our model increased MCC and F1-score by 26.6 and 12.1%, respectively. It is worth noting that GRU performed almost as well as MTAD_GAT, possibly because the multi-head attention mechanism in MTAD_GAT was not as effective for datasets with lower balance, whereas GRU, which focuses solely on temporal features, could outperform it.

In the most imbalanced Dataset C, our model was the only one to achieve an ACC exceeding 70%, while the performance of the MTAD_GAT model was unexpectedly low, as shown in Table 5.4. Since LSTM and GRU are more sensitive to time series, their performance was considerably better than FCN and CNN in this dataset, even surpassing MTAD_GAT. Notably, MTAD_GAT lost its advantage in highly imbalanced datasets. Our model improved the best results achieved by LSTM by nearly 7.1 and 5.3%, respectively, in terms of MCC and F1-score. Compared to the FCN model, which performed the worst, our model increased MCC and F1-score by 20.8 and 7.2%, respectively.

MTAD_GAT's lower performance than expected may be attributed to the inherent challenges in multivariate time-series anomaly detection, such as the influence of noise and missing data, and the difficulty of detecting small and intermittent anomalies. In contrast, our proposed method utilizes the temporal and spatial dependencies of the data through graph neural networks, enabling the detection of subtle changes in wind turbine data.

**Table 5.3** Comparative results on dataset B

| Models | MCC | ACC | F1 | Precision | Recall |
|---|---|---|---|---|---|
| GTAN | **52.4** | **75.7** | **74.5** | **78.1** | **75.7** |
| MTAD_GAT | 45.3 | 72.3 | 70.8 | 74.7 | 72.3 |
| FCN | 25.8 | 63.7 | 62.4 | 63.9 | 63.7 |
| CNN | 36.9 | 67.1 | 63.1 | 73.1 | 67.1 |
| LSTM | 42.4 | 70.7 | 68.8 | 73.7 | 70.7 |
| GRU | 44.8 | 70.7 | 67.8 | 76.8 | 70.7 |

**Table 5.4** Comparative results on dataset C

| Models | MCC | ACC | F1 | Precision | Recall |
|---|---|---|---|---|---|
| GTAN | **44.5** | **70.9** | **68.2** | **76.1** | **70.9** |
| MTAD_GAT | 34.4 | 65.1 | 59.3 | 74.2 | 65.1 |
| FCN | 23.7 | 62.7 | 60.9 | 63.1 | 62.7 |
| CNN | 31.7 | 63.5 | 56.5 | 74.2 | 63.5 |
| LSTM | 37.4 | 67.1 | 62.8 | 74.0 | 67.1 |
| GRU | 36.6 | 66.4 | 61.6 | 74.3 | 66.4 |

### 5.3.3 Ablation Study

In this section, we conducted ablation studies to demonstrate the significance of the Feature Extractor module and Temporal-GAT model within the GTAN framework. We removed these modules from the GTAN model, resulting in GTAN_FE and GTAN_TG, respectively. We then compared the performance of GTAN, GTAN_FE, and GTAN_TG models on datasets A, B, and C using the best hyperparameters selected from $\{1e^{-3}, 1e^{-2}\}$ for the learning rate and [10, 100] for the epoch.

On the balanced Dataset A, both modules contribute to performance enhancement, with the Temporal-GAT model exhibiting a greater impact. As shown in Table 5.5, GTAN_TG reduced the MCC and F1-score by 5.3 and 3.5%, respectively, compared to GTAN, while GTAN_FE reduced the MCC and F1-score by 3.4 and 1.1%, respectively. For the imbalanced Dataset B, Table 5.6 illustrates that both the Feature Extractor module and the Temporal-GAT model have a more pronounced impact, with the Feature Extractor module gradually increasing in importance. GTAN_TG reduced the MCC and F1-score by 9.8 and 8.2%, respectively, compared to GTAN, while GTAN_FE reduced the MCC and F1-score by 7.4 and 4.1%, respectively. On the highly imbalanced Dataset C, the Feature Extractor module and the Temporal-GAT model continue to have a positive influence, with the Feature Extractor module surpassing the Temporal-GAT model in importance for the first time. As illustrated in Table 5.7, GTAN_TG reduced the MCC and F1-score by 2.7 and 4.4%, respectively, compared to GTAN, while GTAN_FE reduced the MCC and F1-score by 7.3 and 6.3%, respectively. The aforementioned results suggest that both the Feature Extractor module and the Temporal-GAT model have a beneficial effect on the model's performance across all datasets.

**Table 5.5** Ablation study on dataset A

| Models | MCC | ACC | F1 | Precision | Recall |
|---|---|---|---|---|---|
| GTAN_TG | 54.0 | 76.4 | 75.3 | 78.9 | 76.4 |
| GTAN_FE | 55.9 | 78.1 | 77.7 | 78.6 | 78.1 |
| GTAN | **59.3** | **79.4** | **78.8** | **80.8** | **79.4** |

**Table 5.6** Ablation study on dataset B

| Models | MCC | ACC | F1 | Precision | Recall |
|---|---|---|---|---|---|
| GTAN_TG | 42.6 | 69.6 | 66.3 | 76.0 | 69.6 |
| GTAN_FE | 45.0 | 72.0 | 70.4 | 74.7 | 72.0 |
| GTAN | **52.4** | **75.7** | **74.5** | **78.1** | **75.7** |

**Table 5.7** Ablation study on dataset C

| Models | MCC | ACC | F1 | Precision | Recall |
|---|---|---|---|---|---|
| GTAN_TG | 41.8 | 68.3 | 63.9 | **77.4** | 68.3 |
| GTAN_FE | 37.2 | 66.7 | 62.0 | 74.5 | 66.7 |
| GTAN | **44.5** | **70.9** | **68.3** | 76.1 | **70.9** |

### *5.3.4 Robustness Analysis*

In this section, we investigate the robustness of the proposed algorithm and validate its performance under varying levels of Gaussian noise through experimentation. We conducted experiments on three basic datasets, A, B, and C, with different levels of Gaussian noise ranging from 0.1 to 0.4, to assess the algorithm's robustness. The algorithm's performance is evaluated by comparing GTAN's performance against that of the MTAD_GAT model, and the experimental results are presented in terms of MCC and F1-score.

Our findings indicate that the proposed algorithm outperforms the comparative algorithm in terms of robustness, as demonstrated in Fig. 5.3. For the balanced Dataset A, both GTAN and MTAD_GAT show decreased performance as the level of Gaussian noise increases. GTAN exhibits a maximum decrease of 1.9 and 0.6% in MCC and F1-score, respectively, while MTAD_GAT shows a maximum decrease of 6.5 and 4.1% in MCC and F1-score, respectively. For the imbalanced Dataset B, both algorithms' performance decreases as the level of Gaussian noise increases, but they also exhibit a rebound phenomenon. GTAN shows a maximum decrease of 7.2 and 5.3% in MCC and F1-score, respectively, while MTAD_GAT shows a maximum decrease of 8.9 and 7.7% in MCC and F1-score, respectively. For the imbalanced Dataset C, GTAN shows a maximum decrease of 7.7 and 6.6% in MCC and F1-score, respectively, while MTAD_GAT shows a maximum decrease of 10.2 and 8.0% in MCC and F1-score, respectively. It is noteworthy that the proposed algorithm almost achieves the best performance of the comparative algorithm under the original data even with high noise levels.

Overall, our experimental results demonstrate that the proposed algorithm possesses superior robustness and can effectively manage time-series data under varying levels of balance.

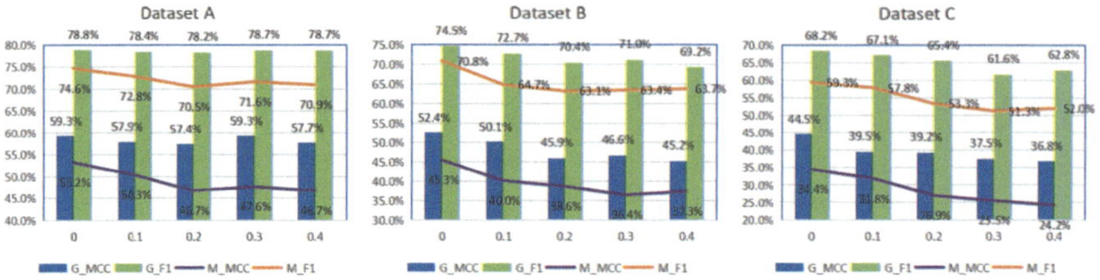


(a) (b) (c)

**Fig. 5.3** Robustness Analysis. Where (**a**), (**b**), (**c**) represent the experimental results on Dataset A, B, and C respectively. G_MCC indicates the MCC value of GTAN, G_F1 indicates the F1-score of GTAN, M_MCC indicates the MCC value of MTAD_GAT, and M_F1 indicates the F1-score of MTAD_GAT. The bottom horizontal coordinate represents the Gaussian noise level

## 5.4 Conclusion

This paper introduces a novel deep learning network to address the challenge of imbalanced sample distribution in wind turbine blade icing detection. The proposed model integrates traditional Graph Attention Networks (GAT) with a Temporal Attention (TA) mechanism, incorporating a Feature Extractor module to manage imbalanced data. While GAT focuses on learning the importance of sensors and time steps, TA addresses the imbalanced data issue. GTAN's competitive performance in imbalanced data scenarios is confirmed through comparisons with five baseline models. The effectiveness of the Feature Extractor module and the Temporal-GAT model is validated through ablation studies and robustness analyses. Notably, the Feature Extractor module demonstrates enhanced performance as the training data becomes more imbalanced. Furthermore, GTAN exhibits significant advantages in robustness compared to the state-of-the-art MTAD_GAT network.

However, the GTAN model has certain limitations that warrant detailed discussion. Firstly, its capability is confined to differentiating between icing and non-icing conditions, without providing information on the severity of ice accumulation. Secondly, the model's training dataset is geographically limited, encompassing wind turbines in close proximity, which raises concerns regarding its universal applicability and generalization across diverse turbine environments and types.

To address these limitations, our future research will prioritize achieving more granular classification of icing severity and will place significant emphasis on improving model performance, particularly in handling imbalanced datasets. Additionally, we will actively seek to incorporate SCADA data from wind turbines situated in diverse environmental contexts and featuring varying installation capacities. Furthermore, we will explore the application of transfer learning techniques to evaluate the GTAN model's performance when applied to wind turbines deployed in distinct geographical regions and operating under varying conditions. This comprehensive approach aims to thoroughly validate the model's generalization capability.

# References


1. U.S. Energy Information Administration, "International Energy Outlook 2023," Energy Information Administration, https://www.eia.gov/outlooks/ieo, Oct 2023 (accessed Sep 2023)
2. Razmjoo A et al (2021) A technical analysis investigating energy sustainability utilizing reliable renewable energy sources to reduce CO2 emissions in a high potential area. Renew Energy 164:46–57
3. Herbert GMJ et al (2007) A review of wind energy technologies. Renew Sustain Energy Rev 11(6):1117–1145
4. Sanchez I (2006) Short-term prediction of wind energy production. Int J Forecast 22(1):43–56
5. Peinke J, Schaumann P, Barth S (eds) (2007) Wind energy: proceedings of the Euromech colloquium. Springer
6. Fortin G, Perron JEAN, Ilinca A (2005) A study of icing events at Murdochville: conclusions for the wind power industry. In: International symposium "Wind Energy in Remote Regions", Magdalen's Island, pp 223–231
7. Laakso T et al (2003) State-of-the-art of wind energy in cold climates. IEA Annex XIX 24:53
8. Wei K et al (2020) A review on ice detection technology and ice elimination technology for wind turbine. Wind Energy 23(3):433–457
9. Madi E et al (2019) A review of integrating ice detection and mitigation for wind turbine blades. Renew Sustain Energy Rev 103:269–281
10. Shu L et al (2018) Numerical model and field experimental investigation of threshold heat flux of wind turbine de-icing. Proc CSEE 38:3997–4003
11. Homola MC, Nicklasson PJ, SundsbÃ PA (2006) Ice sensors for wind turbines. Cold Reg Sci Technol 46(2):125–131
12. Maatuk J. Microprocessor-based liquid sensor and ice detector: U.S. Patent 6,776,037[P]. 2004-8-17
13. Noack RJ. Ice and liquid detector: U.S. Patent 5,796,106[P]. 1998-8-18
14. Zhixin FU, Yue Y (2012) Condition health monitoring of offshore wind turbine based on wireless sensor network. In: 2012 10th international power & energy conference (IPEC). IEEE
15. Kim D-G et al (2017) A fiber-optic ice detection system for large-scale wind turbine blades. Optical Modeling and Performance Predictions IX, vol 10374. SPIE
16. Tao T et al (2021) Wind turbine blade icing diagnosis using hybrid features and Stacked-XGBoost algorithm. Renew Energy 180:1004–1013
17. Guo P, Infield D (2021) Wind turbine blade icing detection with multi-model collaborative monitoring method. Renew Energy 179:1098–1105
18. Mucchielli P et al (2021) Real-time accurate detection of wind turbine downtime-an Irish perspective. Renew Energy 179:1969–1989
19. Hacıefendioğlu K et al (2022) Intelligent ice detection on wind turbine blades using semantic segmentation and class activation map approaches based on deep learning method. Renew Energy 182:1–16
20. Cheng X et al (2022) A blockchain-empowered cluster-based federated learning model for blade icing estimation on iot-enabled wind turbine. IEEE Trans Ind Inform 18(12):9184–9195
21. Zhao H et al (2020) Multivariate time-series anomaly detection via graph attention network. In: 2020 IEEE international conference on data mining (ICDM). IEEE
22. Cheng X et al (2021) Temporal attention convolutional neural network for estimation of icing probability on wind turbine blades. IEEE Trans Ind Electron 69(6):6371–6380
23. Karim F et al (2019) Multivariate LSTM-FCNs for time series classification. Neural Netw 116:237–245
24. Dablain D, Krawczyk B, Chawla NV (2023) DeepSMOTE: fusing deep learning and SMOTE for imbalanced data. IEEE Trans Neural Netw Learn Syst 34(9):6390–6404. https://doi.org/10.1109/TNNLS.2021.3136503. Sept
25. Dosovitskiy A, Springenberg JT, Brox T (2015) Learning to generate chairs with convolutional neural networks. In: Proceedings of the IEEE conference on computer vision and pattern recognition

26. Badrinarayanan V, Kendall A, Cipolla R (2017) Segnet: a deep convolutional encoder-decoder architecture for image segmentation. IEEE Trans Pattern Anal Mach Intell 39(12):2481–2495
27. Ronneberger O, Fischer P, Brox T (2015) U-net: convolutional networks for biomedical image segmentation. In: Medical image computing and computer-assisted intervention–MICCAI 2015: 18th international conference, Munich, Germany, Oct 5–9, 2015, Proceedings, Part III 18. Springer
28. Kim B, Kim J (2020) Adjusting decision boundary for class imbalanced learning. IEEE Access 8:81674–81685
29. Guerriero S et al (2018) DeepNCM: deep nearest class mean classifiers. In: International conference on learning representations
30. Lin T-Y et al (2017) Focal loss for dense object detection. In: Proceedings of the IEEE international conference on computer vision
31. Wang Z, Yan W, Oates T (2017) Time series classification from scratch with deep neural networks: a strong baseline. In: 2017 international joint conference on neural networks (IJCNN). IEEE

# Chapter 6
# Wavelet-Empowered Graph Convolutional Neural Network for Blade Icing Detection

## 6.1 Introduction

The imperative for embracing renewable and ecologically sound energy sources has escalated markedly in recent years, propelled by growing anxieties concerning environmental degradation and the escalating crisis of global warming. Among these clean energy alternatives, wind energy has emerged as a frontrunner, attracting ever-increasing interest and substantial financial commitments. Illustratively, the U.S. Department of Energy has articulated ambitious objectives for wind power generation, aiming to derive 20% of the nation's total electricity from wind turbines by 2030, and further escalating this to 35% by 2050 [1]. Similarly, the European Commission's "Roadmap 2050" initiative envisions wind energy fulfilling half of Europe's electrical demands by mid-century [2]. Realizing these ambitious targets necessitates the deployment of a greater number of turbines in regions characterized by potent wind regimes and frigid, dense air, such as elevated mountainous terrains. Such locations offer wind resources particularly favorable for maximizing wind power production. Globally, installations in cold climates now represent over 30% of the aggregate installed wind turbine capacity [3]. However, these low-temperature environments, especially during winter periods, are notably susceptible to blade icing.

Blade icing is initiated when supercooled water droplets suspended in the atmosphere collide with the surfaces of wind turbine blades. Under conditions of low temperature and diminished liquid water content (LWC), these impinging droplets undergo instantaneous solidification. Conversely, in scenarios characterized by elevated temperatures and higher LWC, only a fraction of the droplets freeze upon impact. The remainder coalesces into a water film across the blade surface, subsequently solidifying [4]. This phenomenon of blade icing introduces significant hazards to both the operational safety and energy efficiency of wind turbines [1]. According to research by Chou et al. [5], failures in turbine blades constitute a primary source of wind turbine incidents, leading to a substantial power reduction of 20–50%, surpassing losses attributed to other adverse factors [5, 6]. The ramifications



X. Cheng et al., *Computational Methods for Blade Icing Detection of Wind Turbines*, Engineering Applications of Computational Methods 24,
https://doi.org/10.1007/978-981-96-6763-5_6

of blade icing are diverse and profoundly impactful, encompassing complete turbine shutdowns, disruptions to aerodynamic performance, overload conditions stemming from delayed aerodynamic stall, diminished fatigue lifespan, and risks to human safety [6]. Analysis of 1,337 turbine standstill events in Sweden between 1998 and 2003, conducted by Jasinski and colleagues [7], revealed that 7% were attributable to blade icing, contributing to an overall power loss of 5%. Investigations by Antikainen et al. [8] and Jasinski et al. [7] have demonstrated that even in nascent stages of icing, imbalances in mass and aerodynamic properties can arise, negatively affecting airfoil effectiveness and the structural integrity of turbines. The Wind Energy Production in Cold Climates (WECO) project further highlighted that ice accumulation can induce elevated deterministic loads and resonance phenomena due to alterations in blade natural frequencies, thereby amplifying fatigue stresses [9]. Moreover, the detachment of ice from rotating blades poses a considerable safety risk, prompting the WECO project to formulate safety distance guidelines for turbine installations [9]. These consequences underscore the critical necessity for a comprehensive understanding of blade icing mechanisms and the development of sophisticated detection methodologies.

Conventional methodologies for detecting icing on wind turbines, primarily blade-centric, rely on physical detectors designed to directly quantify ice properties. These encompass techniques founded on the attenuation of ultrasonic waves [10], thermal infrared radiometry [11], and guided ultrasonic waves [12]. While these direct measurement techniques exhibit reliability, they may exert detrimental effects on blade structural integrity, necessitate considerable financial and labor investments, and often exhibit limited sensitivity to minor ice accumulations [13]. Consequently, practical implementations frequently adopt indirect detection approaches. These methods involve monitoring operational parameters such as power curves [14], anemometer readings [15], vibration time-series data [16], and dew point/temperature conditions [17] to infer icing conditions and initiate de-icing protocols. However, the inherent instability of sensors can introduce inaccuracies in estimations [18], and these indirect methods also encounter challenges in detecting subtle levels of ice accretion [19].

To circumvent the limitations inherent in traditional approaches, research efforts have increasingly gravitated towards data-driven strategies that capitalize on wind turbine monitoring signals. These techniques, broadly classifiable as either model driven or data driven, offer appealing advantages in terms of reduced costs and minimized requirements for mechanical modifications. Model-based methodologies involve constructing mathematical representations, rooted in domain-specific knowledge, that establish correlations between operational signals and icing conditions. These models utilize parameters such as rotor angular velocity [20, 21], power performance curves [22], and thermodynamic principles [23]. However, model-driven approaches are constrained by their reliance on pre-existing domain expertise, limited generalizability across diverse turbine types and environments, and the necessity for costly resources like wind tunnels for parameter optimization. Furthermore, the precision of these models can degrade substantially when operational contexts fluctuate [24].

In contrast, data-driven machine learning techniques, particularly those employing deep learning architectures, have arisen as potent alternatives for blade icing detection. These methods harness the extensive historical datasets available from Supervisory Control and Data Acquisition (SCADA) systems. SCADA systems are ubiquitously deployed for diagnostic evaluations of wind turbine component faults [25, 26], capturing not only operational and environmental metrics (temperature, wind speed, direction) but also delivering a holistic view of turbine status via a network of diverse sensors. The utilization of SCADA data for fault detection presents a cost-effective avenue, obviating the need for supplementary sensor deployments, which has spurred the development of numerous methodologies predicated on SCADA data [25, 27]. These data-centric approaches effectively manage standard sensors, yielding multivariate time-series data that reflect real-time operational environments, ideally suited for machine learning paradigms. Supervised shallow machine learning techniques, including linear discriminant analysis [28], K-Nearest Neighbors (K-NN) [29], Support Vector Machines (SVM) [24], and hybrid methodologies [30–32], have been implemented. For example, Song et al. employed linear discriminant analysis to discern features within power and wind speed distributions, coupled with random forest classifiers for icing detection [28]. Skrimpas et al. utilized K-NN in conjunction with nacelle vibration and power reduction data [29], whereas Xu et al. adopted SVM, optimized via Particle Swarm Optimization (PSO) [24]. Yi et al. introduced MC-SMOTE to mitigate the issue of imbalanced datasets prevalent in fault detection scenarios [30]. However, shallow learning methodologies are significantly dependent on manual feature engineering, a process that is both time intensive and resource demanding.

Deep learning strategies, distinguished by their inherent capacity for automated feature extraction [33], have garnered considerable scholarly interest. Jiang et al. pioneered Multi-scale Convolutional Neural Networks (MSCNN) for wind turbine fault diagnostics [34]. Chen et al. proposed a technique to address the challenges posed by imbalanced data distributions in failure detection contexts [35]. Liu et al. integrated ensemble learning principles into deep learning models to enhance both accuracy and generalization capabilities [36]. Temporal attention convolutional neural networks were engineered by Cheng et al. [37] to discern salient sensors and extract discriminating features from raw operational data. Addressing data imbalance, Cheng et al. also presented a semi-supervised framework for icing detection [38]. Yuan et al. introduced WaveletFCNN, a synthesis of wavelet transformation and fully convolutional neural networks, to derive multi-scale wavelet-based features [39], subsequently refined by Tian et al. [13] through the incorporation of a parallel LSTM-CNN architecture. Wavelet-driven models are particularly pertinent as Discrete Wavelet Transform (DWT) demonstrates efficacy in noise reduction, data compression, and singularity identification, while preserving temporal and spectral information—attributes highly valuable for analyzing non-stationary signals associated with icing, triggered by volatile meteorological factors like wind velocity, direction, and ambient temperature. Percival et al. leveraged DWT for geophysical time-series analysis of Arctic Sea ice [40], and Nalley et al. employed DWT in conjunction with the Mann-Kendall test for trend analysis in stream flow and precipitation patterns [41]. Nobre et al. combined PCA, DWT, and XGBoost for forecasting market direction [42]. Despite advancements in wavelet-based approaches for blade icing detection, including WaveletFCNN [39], WaveletAE [43], and MCRNN [13], current methodologies often treat wavelet-transformed data as direct inputs, overlooking the comprehensive exploration of inter-variable correlations across diverse wavelet scales. These interdependencies, spanning distinct temporal and frequency domains, are critical for achieving accurate representations in complex, real-world operational scenarios [44, 45]. Furthermore, the intrinsic data imbalance within SCADA datasets, where icing incidents are infrequent relative to normal operation, presents a considerable obstacle to the effective training of data-driven detection models.

To surmount these salient challenges, this paper introduces a wavelet-driven multi-scale graph convolutional network (MWGCN) model specifically designed for wind turbine blade icing detection. This innovative model capitalizes on multi-scale discrete wavelet decomposition to extract salient features across both temporal and spectral domains, and subsequently applies scale-specific correlation learning to automatically infer scale-dependent graph structures. This enables the model to proficiently capture implicit inter-variable relationships both within and across varied time and frequency scales. Specifically, the MWGCN architecture incorporates a temporal graph convolutional network, integrating a graph convolutional network (GCN) with a temporal convolutional network (TCN), to process multi-scale information, moving beyond conventional RNN-based paradigms. To further refine the model's performance, a scale-attention module is integrated to modulate the significance of temporal patterns at different scales and to capture inter-scale dependencies. Finally, to alleviate the impact of data imbalance, a pervasive characteristic of wind farm SCADA data, a class rebalancing classifier, utilizing focal loss, is proposed to enhance classification efficacy.

The principal contributions of this research are delineated as follows:

1. We propose a novel wavelet-driven multi-scale graph convolutional network (MWGCN) framework for the detection of icing on wind turbine blades. To the best of our knowledge, this work represents the pioneering application of a graph convolutional model within the domain of wind turbine blade icing detection.
2. We develop a unique network architecture meticulously crafted to model the interrelations of multi-scale wavelets in both the time and frequency dimensions, thereby facilitating superior extraction of features pertaining to inter-variable correlations across multiple wavelet scales.
3. We undertake exhaustive experimentation utilizing real-world SCADA data acquired from wind farms to rigorously evaluate the proposed model, empirically demonstrating its effectiveness, superiority, and enhanced performance relative to a suite of eight baseline methodologies.

The subsequent sections of this paper are organized as follows: Sect. 6.2 provides a detailed exposition of the proposed model architecture; Sect. 6.3 presents and critically analyzes the experimental findings and comparative evaluations against baseline approaches; and Sect. 6.4 concludes the paper, offering concluding remarks and outlining prospective avenues for future investigation.

## 6.2 Methodology

This section will first present the overall framework of the proposed Multi-scale Wavelet Graph Convolutional Network (MWGCN), then describe all modules in greater detail, including the data preprocessing module, the multi-scale wavelet decomposition module, the scale-specific correlation learning module, the Temporal Graph Convolutional (TGC) module, the scale-attention module, and the blade icing classifier.

### *6.2.1 Overview*

Figure 6.1 provides a visual representation of the comprehensive framework implemented for blade icing detection in wind turbines, which is divided into several key stages: (a) historical raw data collection via the Supervisory Control and Data Acquisition (SCADA) system of the wind turbine, forming the basis for subsequent analysis; (b) data preprocessing to refine the raw data; (c) decomposition of preprocessed data into multiple scales using the Multi-scale Discrete Wavelet Decomposition (MDWD) module to extract multiresolution features; (d) analysis of inter-variable relationships across various wavelet scales by the scale-specific correlation learning module; (e) generation of scale-specific features by the Graph Convolutional Network (GCN) submodule, followed by feature extraction from each wavelet detail coefficient within a defined window size using the Temporal Convolutional Network (TCN) submodule; (f) refinement of scale-specific features through the scale-attention module to emphasize pertinent information; (g) final classification performed by the class rebalancing classifier to determine icing conditions; and (h) integration of the fully trained model with SCADA systems to enable real-time, online blade icing detection.

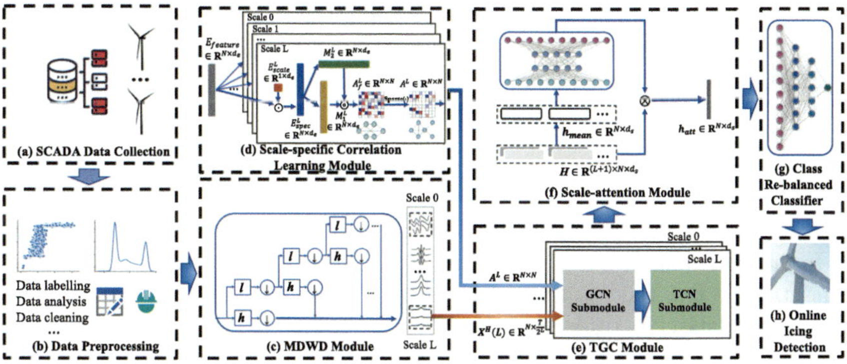


**Fig. 6.1** The overall framework of blade icing detection model MWGCN for wind turbines

### 6.2.2 Data Preprocessing

In this study, multivariate time-series data are sourced from a Supervisory Control and Data Acquisition (SCADA) system, which is outfitted with general-purpose sensors monitoring crucial operational parameters such as wind speed, generator speed, power output, wind direction, and yaw speed. However, the reliability of these sensors can be compromised by mechanical issues or human errors, potentially leading to data recording interruptions or the generation of anomalous readings. Consequently, the implementation of a robust data preprocessing module is paramount to mitigate the influence of outliers and address data gaps. This module encompasses several essential functions: data labeling to categorize observations, data normalization to ensure uniformity across variables, statistical analysis to identify and rectify anomalies, and effective strategies for managing missing data entries. This holistic approach ensures the integrity and dependability of the data before further analytical processing.

**(1) Data labeling**

Experienced wind turbine engineers were tasked with the critical process of labeling the raw blade icing data. Recognizing the inherent operational variability of wind turbines, which often complicates the differentiation between normal and icing conditions even for seasoned experts, certain ambiguous data intervals were intentionally excluded from labeling. Within a defined time step, known as the window size, engineers categorized the data into two distinct classes: "normal" for standard operational scenarios and "blade icing" to denote periods affected by icing. This meticulous labeling procedure is fundamental for training accurate models capable of detecting and predicting blade icing based on operational data patterns.

**(2) Data Normalization**

The multivariate time-series data derived from SCADA systems frequently contain substantial noise, making normalization of raw data essential for effective noise reduction. Normalization not only improves the efficiency of the training process but also accelerates the convergence rate of the learning models. Therefore, this research adopts the Min-Max normalization technique to preprocess the data. For a given SCADA signal represented as a dataset $X = (x_1, x_2, ..., x_T)$, where $T$ is the number of data points in the time series, the normalization of $X$ is achieved using the following formula:

$$x_{i,\text{norm}} = \frac{x_i - x_{\min}}{x_{\max} - x_{\min}}. \tag{6.1}$$

In this equation, $X_{\text{norm}} = (x_{1,\text{norm}}, x_{2,\text{norm}}, ..., x_{T,\text{norm}})$ represents the normalized dataset, while $x_{\max}$ and $x_{\min}$ are the maximum and minimum values observed within the original dataset $X$, respectively.

**(3) Statistics Analysis**

The SCADA system in wind turbines collects data from a multitude of sensors; however, a significant portion of this data may be redundant, potentially hindering the performance of detection models and increasing computational costs. To address this issue, a comprehensive statistical analysis of SCADA data is crucial to optimize the blade icing detection process. This study employs two key statistical techniques to streamline the data:

1. Identification and Removal of Redundant Features: To identify and eliminate features that contribute minimally or not at all to blade icing detection, we utilize the Pearson correlation coefficient. This method aids in assessing the linear relationship between variables, enabling the removal of redundant data that does not significantly enhance model performance.
2. Mitigation of Multicollinearity: To mitigate the risk of severe overfitting in detection models, which can arise due to multicollinearity among features, we address this issue by replacing highly correlated features with their average values. This approach simplifies the model without sacrificing critical information, thereby improving the robustness and generalizability of the detection model.

**(4) Missing Data Processing**

Due to occasional sensor malfunctions or communication disruptions, the SCADA system may not always acquire complete datasets. To effectively manage gaps in data collection, this study employs two methodologies for handling missing values, based on the duration of data absence:

1. Short-duration Missing Values: Linear interpolation is applied when data points are missing for short intervals, such as a few consecutive readings. This method estimates missing values based on the data points immediately preceding and succeeding the gap.
2. Long-duration Missing Values: For prolonged periods of missing data, such as several hours or days, historical data from analogous time periods in the past are used to fill these gaps. This approach helps to avoid potential inaccuracies that interpolation might introduce over extended durations.

### *6.2.3 Multi-scale Wavelet Decomposition*

Given the intricate nature of wind turbines and the variable conditions under which they operate, the time-series data captured by SCADA systems are inherently complex and fluctuating [13]. To effectively capture and analyze these complexities, wavelet decomposition is utilized to examine the data across both time and frequency domains, thereby revealing deeper insights. The multi-scale wavelet decomposition module employs the Multilevel Discrete Wavelet Decomposition (MDWD) algorithm [46] to extract features at multiple scales from an input time series. This includes both low- and high-frequency subseries, as well as detailed scale-by-scale analysis, highlighting signal variations at different resolutions. The operational framework of the MDWD module is illustrated in Fig. 6.1c.

Within the Multi-scale Discrete Wavelet Decomposition (MDWD) module, each input signal segment from the SCADA dataset is denoted as $x_c = [x_{c,1}, x_{c,2}, ..., x_{c,T}] \in \mathbb{R}^T$, where $c = 1, 2, 3, ..., N$ indicates the signal index, and $T$ represents the length of the input segment (window size). Consequently, the matrix $X = [x_1, x_2, ..., x_N] \in \mathbb{R}^{N \times T}$ encapsulates the input multivariate time-series segments. The low and high subseries produced at the $i$-th scale are represented by $x^l_{(i)}$ and $x^h_{(i)}$, respectively. At the $(i+1)$-th scale, the application of a low-pass filter $l = [l_1, l_2, ..., l_K]$ and a high-pass filter $h = [h_1, h_2, ..., h_K]$ $(K \ll T)$ results in the convolution of the approximate coefficients (low-frequency subseries) from the previous scale:

$$
\begin{aligned}
a^l_n(i+1) &= \sum_{k=1}^{K} x^l_{n+k-1}(i) \cdot l_k, \\
a^h_n(i+1) &= \sum_{k=1}^{K} x^l_{n+k-1}(i) \cdot h_k,
\end{aligned}
\tag{6.2}
$$

where $x^l_n(i)$ is the $n$-th element of the approximate coefficients at the $i$-th scale. The process is iterated up to $L$ MDWD levels, with $L$ being the number of decomposition stages. The new approximate coefficients $x^l(i+1)$ and the detailed coefficients (high-frequency subseries) $x^h(i+1)$ at scale $(i+1)$ are derived by down-sampling the intermediate variables $a^l(i+1) = [a^l_1(i+1), a^l_2(i+1), ...]$ and $a^h(i+1) = [a^h_1(i+1), a^h_2(i+1), ...]$ by a factor of 1/2. Each signal's concatenated segment can then be represented as $X_c = [x_c, x^h(1), ..., x^h(L)]$, and, accordingly, the complete input dataset for the model is $X_{\text{input}} = [X, X^H(1), ..., X^H(L)]$. For simplification, the original multivariate time-series data is designated as the wavelet detail coefficient at the 0-th scale, $X^H(0)$.

### 6.2.4 Scale-Specific Correlation Learning

Following the acquisition of wavelet detail coefficients for input signals across multiple scales, a scale-specific correlation learning module is implemented to automatically identify inter-variable relationships that are unique to each scale. This module is inspired by advancements in network science as detailed in [47]. In contrast to conventional wavelet-based models [13, 39], which typically do not emphasize scale-specific inter-variable correlations, our approach acknowledges the necessity of capturing diverse patterns across different time and frequency domain scales. Consequently, the ability to learn these multi-scale-specific inter-variable correlations becomes essential. The architecture of this module is depicted in Fig. 6.1d.

Constructing a separate adjacency matrix for each scale can be computationally demanding and may introduce excessive noise. To mitigate these issues, we construct the adjacency matrix using two types of embedding vectors: the shared feature embedding $E_{\text{feature}} \in \mathbb{R}^{N \times d_e}$ and a unique scale embedding for each scale $E^i_{\text{scale}} \in \mathbb{R}^{1 \times d_e}$, where $d_e$ is the embedding dimension. These embeddings, containing both shared and scale-specific information, are initially randomized and subsequently refined through the training process. The adjacency matrix at the $i$-th scale is computed as follows:

$$
\begin{aligned}
E^i_{\text{spec}} &= E_{\text{feature}} \odot E^i_{\text{scale}}, \\
M^i_1 &= \left[\tanh\left(E^i_{\text{spec}} \theta^i\right)\right]^T, \\
M^i_2 &= \tanh\left(E^i_{\text{spec}} \varphi^i\right), \\
A^i_{\text{full}} &= ReLU\left(M^i_1 M^i_2 - \left(M^i_1 M^i_2\right)^T\right), \\
A^i &= Sparse\left(Softmax\left(A^i_{\text{full}}\right)\right),
\end{aligned}
\tag{6.3}
$$

here, $E^i_{\text{spec}}$ denotes the embedding of the adjacency matrix for the $i$-th scale, modulated by two trainable parameters $\theta^i$ and $\varphi^i$ to generate matrix factorization vectors $M^i_1$ and $M^i_2$. The raw adjacency matrix $A^i_{\text{full}}$ is then processed to create a sparse matrix:

$$
Sparse(A^i_{lr}) = \begin{cases} A^i_{lr}, & \text{if } A^i_{lr} \in Top\left(A^i_{l*}, \tau\right) \\ 0, & \text{if } A^i_{lr} \notin Top\left(A^i_{l*}, \tau\right), \end{cases} \tag{6.4}
$$

where $\tau$ is a threshold that defines the number of neighbors each node retains in the graph. This methodology effectively constructs the adjacency matrix for each scale, denoted as $A = [A^0, A^1, A^2, ..., A^L]$, illustrating the scale-specific inter-variable relationships across different wavelet scales.

### 6.2.5 *Temporal Graph Convolutional Network*

The temporal graph convolutional module incorporates $L$ Temporal Convolutional Networks (TCNs) and Graph Convolutional Networks (GCNs) for each scale. It utilizes the adjacency matrix set $A = [A^0, A^1, A^2, ..., A^L]$ generated by the scale-specific correlation learning module and the multi-scale wavelet input $X_{\text{input}} = [X, X^H(1), ..., X^H(L)]$ from the MDWD module. This module differs from RNN-like modules, such as LSTM and GRU, which possess inherent delays between sequential time steps. The TCN submodule's parallel computing capability enhances the model's efficiency during both training and testing phases.

The processing within this module at each time step begins with the GCN submodule, which prepares the inputs for the TCN submodule at different scales:

$$h_t^i = GCN_{\text{in}}^i(x_t^i, A^i) + GCN_{\text{out}}^i(x_t^i, (A^i)^T), \tag{6.5}$$

where $h_t^i$ represents the input to the $i$-th scale's TCN at time $t$, and $x_t^i$ is the wavelet input at the $i$-th scale for the same time step. The operation within the GCN is described by the graph convolution formula:

$$x *_G \theta = \sigma\left(\theta\left(\widetilde{D}^{-\frac{1}{2}}\widetilde{A}\widetilde{D}^{-\frac{1}{2}}\right)x\right), \tag{6.6}$$

here, $G$ represents the graph defined by the adjacency matrix $A$, $x$ indicates the nodes' representations containing the $N$ signals for the specific wavelet scale, $\sigma$ denotes the activation function, $\theta$ is a trainable parameter matrix, $\widetilde{A} = I_n + A$ is the augmented adjacency matrix including self-connections, and $\widetilde{D}$ is the diagonal degree matrix of $\widetilde{A}$.

Subsequently, the outputs of the GCN submodule for the $i$-th scale, $\left\{h_1^i, h_2^i, ..., h_{\frac{T}{2^i}}^i\right\}$, are passed to the TCN submodule to derive the scale-specific representations $h^i$:

$$h^i = TCN^i(h_1^i, h_2^i, ..., h_{\frac{T}{2^i}}^i), \tag{6.7}$$

where $TCN^i(\cdot)$ refers to the TCN processing the wavelet detail coefficients at the $i$-th scale, and the dimensionality of $h^i$ is set to $d_s$. This process effectively captures and synthesizes time-dependent and scale-specific features for subsequent analysis or prediction tasks.

### 6.2.6 Scale-Attention Module

The scale-specific features $H = [h^0, h^1, h^2, ..., h^L] \in \mathbb{R}^{(L+1)\times N\times d_s}$ are processed in the scale-attention module to evaluate the relevance of temporal patterns across different wavelet scales and to identify cross-scale correlations. This module initiates by performing an average pooling operation along the scale dimension to consolidate scale-specific information across the $N$ signal variables, thus revealing potential cross-scale correlations:

$$h_{mean}^j = \frac{\sum_{i=0}^{L} H_j^i}{L+1}. \tag{6.8}$$

In this equation, $h_{\text{mean}}^j \in \mathbb{R}^{1\times d_s}$ represents the aggregated information for the signal variable $j = 1, 2, ..., N$, forming a global representation $h_{\text{mean}} \in \mathbb{R}^{N\times d_s}$.

Subsequently, to compute the attention weights for each scale, a sequence of fully connected layers is utilized:

$$
\begin{aligned}
a_{\text{in}} &= ReLU(\mathbf{W_{in}} h_{\text{mean}} + b_{\text{in}}), \\
a_{\text{out}} &= Sigmoid(\mathbf{W_{out}} a_{\text{in}} + b_{\text{out}}),
\end{aligned} \tag{6.9}
$$

where $a_{\text{out}} \in \mathbb{R}^{1\times(L+1)}$ specifies the attention weights for the scale-specific features $H$. Here, $\mathbf{W_{in}}$ and $\mathbf{W_{out}}$ are the weight matrices for the fully connected layers, and $b_{\text{in}}$ and $b_{\text{out}}$ are the corresponding biases. The final scale-attention representation $h_{\text{att}} \in \mathbb{R}^{N\times d_s}$ is then calculated by applying these weights to the features:

$$
h_{\text{att}} = ReLU(a_{\text{out}} \otimes H). \tag{6.10}
$$

## 6.2.7 Class Rebalancing Blade Icing Classifier

The scale-attention representation $h_{\text{att}}$ is fed into a class rebalancing classifier for the final blade icing detection. This classifier consists of a multilayer fully connected network employing ReLU and Sigmoid activation functions to process the data across multiple layers:

$$
\begin{aligned}
o_0 &= \text{ReLU}(\mathbf{W_0} h_{\text{att}} + b_0), \\
o_{i+1} &= \text{ReLU}(\mathbf{W}_i o_i + b_i) \quad \text{for} \quad i = 0, 1, ..., f-1, \\
\hat{y} &= \text{Sigmoid}(\mathbf{W}_f o_f + b_f),
\end{aligned} \tag{6.11}
$$

where $o_0$ represents the output from the initial layer; $o_{i+1}$ is the output from each subsequent layer; and $\hat{y}$ is the final prediction, indicating the probability of blade icing, normalized to a range between 0 and 1.

Given the inherent imbalance in SCADA data, where non-icing conditions significantly outnumber icing events, we utilize the focal loss function to enhance class balance by prioritizing hard-to-classify samples. The focal loss function extends the conventional cross-entropy loss by incorporating a modulating factor that reduces the influence of well-classified (easy) examples and amplifies the contribution from more challenging ones, thereby improving performance on minority classes:

$$
\text{FL}(p_t) = -\alpha_t (1 - p_t)^{\gamma} \log(p_t), \tag{6.12}
$$

where $p_t$ is the predicted probability for the true class label (blade icing), $\alpha_t$ balances the importance between positive and negative classes, and $\gamma$ adjusts the focus towards harder examples. This adjustment helps to counteract bias towards the majority class and enhances the model's capacity to accurately detect blade icing events.

## 6.3 Experiment

This section details the experimental assessment of the proposed methodology, encompassing the experimental configurations, data preprocessing steps, evaluation metrics, implementation specifics, comparisons with baseline models, ablation study, and sensitivity analysis.

### *6.3.1 Experimental Setups*

**(1) Implementation Settings**

The Pytorch deep learning framework (v1.10.2) was used for the implementation of the neural networks in the proposed model. All experiments were conducted on an Ubuntu (v.18.04.5) server, featuring a 2.20 GHz Intel Xeon CPU, NVIDIA Tesla T4 GPU, and 16 GB of RAM. The Adam optimizer was employed during model training [48]. We maintained parameter settings consistent with those of the baseline models, without hyperparameter tuning.

**(2) Datasets**

The experimental evaluation utilized SCADA data from two wind turbines situated in the Inner Mongolia Autonomous Region, China, provided by the Chinese Ministry of Industry and Information Technology (MIIT).[1] Data acquisition for each turbine was performed at 7-second intervals, resulting in total operational durations of 696 and 306 h for Turbine-1 and Turbine-2, respectively. Across each turbine, over a hundred sensors record various operational metrics; however, this analysis focuses on 26 signals pertinent to blade icing conditions, as identified by experienced wind turbine engineers. Table 6.1 provides a comprehensive description of these datasets, which will be referred to as Turbine-1 and Turbine-2 hereafter.

The data processing methodology adopted in this study is consistent with approaches documented in relevant literature [13]. As detailed in Table 6.2, wind turbines predominantly operate under normal conditions, with blade icing events occurring only sporadically. This results in a significant imbalance between "normal" and "icing" labels. Furthermore, the dataset includes entries lacking labels, likely due to unidentified errors. The initial phase of our data processing involves cleaning these unlabeled data entries.

Additionally, the SCADA signals are segmented into fixed time steps to serve as inputs for the model. This fixed-length time step is consistently referred to as the "window size". Each segmented sample is then assigned a binary label, indicating the presence or absence of blade icing during that specific window.

[1] http://106.38.3.217/pages/datadown/dataset/dataSet06.jsp.

**Table 6.1** Description of the datasets

| No. | Parameter | No. | Parameter |
|---|---|---|---|
| 1 | Wind speed | 14 | Temperature of pitch motor 1 |
| 2 | Generator speed | 15 | Temperature of pitch motor 2 |
| 3 | Active power | 16 | Temperature of pitch motor 3 |
| 4 | Live wind direction | 17 | Switching temperature of pitch 1 |
| 5 | Nacelle temperature | 18 | Switching temperature of pitch 2 |
| 6 | Yaw position | 19 | Switching temperature of pitch 3 |
| 7 | Yaw speed | 20 | Charger's DC current of pitch 1 |
| 8 | Angle of pitch 1 | 21 | Charger's DC current of pitch 2 |
| 9 | Angle of pitch 2 | 22 | Charger's DC current of pitch 3 |
| 10 | Angle of pitch 3 | 23 | Horizontal acceleration |
| 11 | Speed of pitch 1 | 24 | Vertical acceleration |
| 12 | Speed of pitch 2 | 25 | Environment temperature |
| 13 | Speed of pitch 3 | 26 | Average wind direction with 25s |

**Table 6.2** Label distribution of the datasets

| Dataset | Total | Normal | Icing | Invalid |
|---|---|---|---|---|
| Turbine-1 | 393,886 | 350,255 | 23,846 | 19,785 |
| Turbine-2 | 190,494 | 168,930 | 10,638 | 10,926 |

### (3) Metrics

Wind turbines typically operate under normal conditions for the majority of their operational lifespan, with blade icing being an infrequent occurrence. This operational characteristic leads to a higher prevalence of negative samples (normal states) compared to positive samples (icing states) in SCADA datasets, resulting in a notable class imbalance. Specifically, the ratio of negative to positive samples is approximately 15:1, highlighting the challenge of managing skewed class distributions in the analysis.

In this study, precision is defined as the proportion of correctly identified blade icing incidents among all instances classified as positive, while recall quantifies the proportion of actual icing incidents that are correctly identified. A high incidence of false positives can result in unnecessary manual inspections and maintenance, leading to increased costs and operational inefficiencies. Precision effectively captures this aspect, with higher precision values indicating fewer false positives. Conversely, a high rate of false negatives can lead to undetected icing conditions, potentially resulting in turbine failures. Recall quantifies this risk, with higher recall values corresponding to fewer missed icing events. However, in datasets with a significant class imbalance in blade icing instances, high precision alone may not accurately reflect model performance. In such scenarios, models might only classify the most obvious icing events, overlooking more subtle cases and thus exhibiting low recall despite high precision. Conversely, a model optimized for high recall might over-classify instances as icing, leading to many false positives and unnecessary maintenance. This trade-off underscores the challenge in balancing precision and recall, particularly in imbalanced datasets where both metrics are crucial for a comprehensive assessment of model effectiveness.

Therefore, addressing the imbalance in blade icing datasets necessitates a metric that provides equal consideration to both precision and recall, offering a more balanced evaluation of model performance. Drawing from established literature [13, 39, 49], the F1-score is advocated for such scenarios. The F1-score, representing the harmonic mean of precision and recall, serves as a robust measure that ensures neither precision nor recall is disproportionately emphasized at the expense of the other. Consequently, this study calculates precision and recall, alongside the F1-score, to evaluate model efficacy, with the F1-score serving as the primary metric for assessing experimental outcomes. This approach enables a more thorough evaluation, especially critical in imbalanced scenarios typical of blade icing detection datasets.

The outcomes of blade icing detection are effectively summarized using a confusion matrix, a $2 \times 2$ matrix that categorizes predictions into four categories: True Positive (TP), True Negative (TN), False Positive (FP), and False Negative (FN). Within this matrix, samples correctly identified as exhibiting icing conditions are counted as True Positives (TP), and samples correctly identified as normal are True Negatives (TN). Conversely, samples incorrectly predicted as having icing conditions are False Positives (FP), and those with undetected icing conditions are False Negatives (FN). This matrix offers a clear visualization of model performance by distinguishing between correctly and incorrectly classified samples, with samples labeled as experiencing icing conditions designated as positive, and normal operational samples as negative.

The evaluation metrics for blade icing detection models are then computed based on TP, TN, FP, and FN, including precision, recall, and the F1-score, as described by the following equations:

$$\begin{aligned} Precision &= \frac{TP}{TP+FP}, \quad Recall = \frac{TP}{TP+FN}, \\ F1-score &= \frac{2 \times Precision \times Recall}{Precision + Recall}. \end{aligned} \tag{6.13}$$

Given the dataset's imbalance, accuracy is not employed as a primary performance metric in this study. For instance, a model that classifies all samples as normal might achieve high accuracy by correctly identifying the majority of normal samples, but it would fail to detect icing events, rendering accuracy a misleading indicator of performance under class imbalance. Similarly, specificity (the True Negative Rate, TNR) and the False Positive Rate (FPR), where $FPR = 1 - TNR$, are also not utilized as metrics for analogous reasons. These measures could spuriously suggest high performance while inadequately capturing the model's ability to correctly identify the less frequent icing events.

**(4) Model Settings**

The window size for data preprocessing is set at 32. In the multi-scale wavelet decomposition module, the input multivariate time-series segments undergo decomposition at three scales using the Haar wavelet transform. For the scale-specific correlation learning module, the embedding size, denoted as $d_e$, is set to 4, with the maximum number of neighbors for each node ($\tau$) capped at 10. The Graph Convolutional Network (GCN) submodule comprises two layers of graph convolution, while the Temporal Convolutional Network (TCN) submodule utilizes convolution layers with kernel sizes of 32, 16, 8, and 4 across the different scales, respectively. The class rebalancing blade icing classifier is structured with three fully connected layers, consisting of 26, 13, and one node(s), respectively. The parameters for the focal loss function are specifically configured with $\alpha_t$ at 0.25 and $\gamma$ at 3, optimized for handling class imbalance. A detailed sensitivity analysis of these and other crucial hyperparameters is presented subsequently in Sect. 6.3.4.

## *6.3.2 Comparison with the Baseline Methods*

We employ the following eight state-of-the-art models as baselines to evaluate our proposed model.

**MSPCA** is a moving window generalized likelihood ratio test method grounded in multi-scale principal component analysis, utilizing wavelet-based multi-scale data representation [50]. **RF-XGBoost** is a wind turbine fault detection methodology that integrates random forest and XGBoost within an ensemble learning framework [51]. **LSTM** is an artificial recurrent neural network model specifically engineered to address long-term dependencies in sequential data. **MLSTM-FCN** is an efficient multivariate time-series classification model known for its performance and computational efficiency [52]. **FCNN** is a straightforward yet competitive baseline for time-series classification, employing deep neural networks from scratch [53]. **WaveletFCNN** is a wavelet-based fully convolutional neural network designed for wind turbine blade icing detection, leveraging wavelet features [39]. **SSENET** is constructed upon stacked convolutional neural network (CNN) blocks, incorporating dense connections and channel and feature attention modules to enhance feature extraction [33]. **MRCNN** is a wavelet-based multilevel convolutional recurrent neural network that combines wavelet analysis with deep learning for improved time-series modeling [13].

Table 6.3 presents a comparative analysis of the performance of the proposed Multi-scale Wavelet Graph Convolutional Network (MWGCN) and various baseline models on the Turbine-1 dataset, with a focus on precision, recall, and F1-score. The results clearly indicate that the MWGCN outperforms all other models in terms of precision and the primary evaluation metric, the F1-score. Specifically, in comparison to the most competitive baseline, the Multilevel Convolutional Recurrent Neural

**Table 6.3** Baseline comparison on the datasets

| Dataset | Turbine-1 | | | Turbine-2 | | |
|---|---|---|---|---|---|---|
| | Precision | Recall | F1-score | Precision | Recall | F1-score |
| MSPCA | 33 | 30 | 31 | 30 | 66 | 41 |
| RF-XGBoost | 33 | 79 | 47 | 31 | 71 | 43 |
| LSTM | 27 | 57 | 37 | 32 | 30 | 31 |
| MLSTM-FCN | 37 | 78 | 50 | 37 | 95 | 53 |
| FCNN | 26 | 82 | 39 | 38 | 73 | 50 |
| WaveletFCNN | 21 | 60 | 31 | 19 | 84 | 31 |
| SSENET | 53 | 59 | 56 | 35 | 90 | 50 |
| MRCNN | 53 | 81 | 64 | 69 | 72 | 71 |
| MWGCN | **76** | 74 | **75** | **88** | 72 | **79** |

Network (MRCNN), the MWGCN demonstrates an improvement of 17.2% in the F1-score and 43.4% in precision. While certain models, such as FCNN, MRCNN, and Multivariate LSTM-Fully Convolutional Network (MLSTM-FCN), exhibit higher recall than the MWGCN, their precision metrics are notably lower. High recall coupled with low precision can result in a substantial number of false positives. In practical scenarios, this translates to these models being more likely to erroneously classify normal operational periods as blade icing events, which can trigger unnecessary maintenance interventions, leading to significant additional expenses. Thus, while recall is an important performance indicator, the substantial precision offered by the MWGCN enhances its practical utility by minimizing the occurrence of costly false alarms.

Table 6.3 also details the blade icing detection performance of the Multi-scale Wavelet Graph Convolutional Network (MWGCN) and the baseline models on the Turbine-2 dataset, focusing on F1-score and overall accuracy. The results again demonstrate the superiority of the MWGCN, which outperforms the best baseline model, the Multilevel Convolutional Recurrent Neural Network (MRCNN), by achieving improvements of 11.3% in the F1-score and 27.5% in accuracy. Although some models generate false positives, indicating a need for enhanced balance between precision and recall, the proposed MWGCN exhibits superior capability in managing these aspects. This balanced approach enables the MWGCN to attain the highest F1-scores across both datasets. The ability to maintain high precision while also achieving a commendable recall rate positions the MWGCN as an effective solution for minimizing erroneous operational disruptions and unnecessary maintenance costs associated with misclassified icing events.

### 6.3.3 Computational Complexity

During the testing phase of the Multi-scale Wavelet Graph Convolutional Network (MWGCN), the computationally intensive modules include the Multi-scale Discrete Wavelet Decomposition (MDWD) module, the Graph Convolutional Network (GCN) submodule, the Temporal Convolutional Network (TCN) submodule, the scale-attention module, and the class rebalanced classifier. Their respective computational complexities are: $O(N \cdot T \cdot L)$ for the MDWD module, $O(N^2 \cdot T \cdot L)$ for the GCN submodule, $O(T^2 \cdot N \cdot d_s \cdot L)$ for the TCN submodule, and $O(d_s^2)$ for both the scale-attention module and the class rebalanced classifier, where $N$ is the number of signals, $T$ is the window size, $L$ is the number of MDWD levels, and $d_s$ is the embedding size of $h^i$ and $h_{\text{att}}$, respectively.

### 6.3.4 Sensitivity Analysis and Hyperparameter Tuning

This section presents a sensitivity analysis to evaluate the influence of critical hyperparameters on the performance of the Multi-scale Wavelet Graph Convolutional Network (MWGCN), aiming to identify optimal configurations and refine the network architecture.

We concentrate on three key hyperparameters, utilizing data from the Turbine-1 and Turbine-2 datasets to guide our investigation:

1. MDWD Scale: The number of scales employed in the Multi-scale Discrete Wavelet Decomposition (MDWD) module determines the extent of data decomposition and the granularity of extracted information.
2. Maximum Neighbor Number: This hyperparameter in the scale-specific correlation learning module dictates the maximum count of neighboring nodes each node can interact with, affecting the network's capacity to learn intricate inter-node relationships.
3. Number of Graph Convolution Layers: The depth of the GCN submodule, indicated by the number of graph convolution layers, governs the complexity and capability of the model to capture hierarchical features within the data.

The ranges for these hyperparameters were defined based on recommendations from relevant literature [13] and empirical insights derived from practical experimentation. This analysis ensures that the MWGCN is not only theoretically robust but also optimized for effective performance under real-world conditions, leading to more accurate and reliable blade icing detection.

To assess the impact of the MDWD scale on the MWGCN's performance, we tested four distinct scales (1, 2, 3, and 4). We also examined the influence of the maximum number of neighbors per node in the scale-specific correlation learning module by varying it from 5 to 26 across five settings. Additionally, we evaluated the effect of varying the number of graph convolution layers in the GCN submodule,

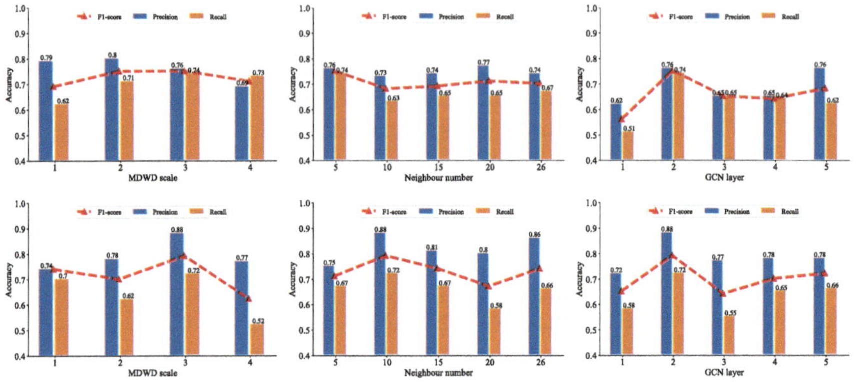


**Fig. 6.2** Results of the sensitivity analysis on the Turbine-1 dataset (upper line) and the Turbine-2 dataset (lower line)

testing configurations with 1, 2, 3, 4, and 5 layers. In total, this resulted in 14 sets of sensitivity experiments conducted on the Turbine-1 and Turbine-2 datasets.

The results of the sensitivity analysis for both datasets are depicted in Fig. 6.2. For the Turbine-1 dataset, optimal F1-scores were achieved with MDWD scales of 2 and 3. Given the potential economic consequences of undetected blade icing events, a scale of 3 was selected due to its superior recall compared to scale 2. Furthermore, the highest F1-score was observed when the number of neighbors was set to 5, and model performance peaked with two layers in the GCN submodule. For the Turbine-2 dataset, similar trends were observed concerning the MDWD scale and the number of GCN layers, with the most favorable F1-score obtained when the number of neighbors was increased to 10. These findings facilitated the determination of the most effective hyperparameter settings for the MWGCN on each dataset, optimizing model accuracy and reliability in blade icing detection.

### *6.3.5 Ablation Study*

An ablation study is conducted to rigorously evaluate the effectiveness and necessity of specific components within a model. This methodology involves systematically removing individual elements from the model to observe the resulting impact on performance. By comparing the performance metrics of the model with and without each component, it is possible to ascertain the contribution of that specific element to the overall functionality of the model. For example, if a novel component is integrated into an existing model, the ablation study would assess the model's performance both without and with this addition. This comparative analysis helps in determining whether the new component enhances the model's effectiveness or if it is redundant.

To thoroughly assess the individual contributions of various modules within the Multi-scale Wavelet Graph Convolutional Network (MWGCN), we performed an ablation study by creating several variants of the model, each lacking a specific module. This approach aids in understanding the significance of each component in enhancing the comprehensive performance of the MWGCN.

The variants constructed are described as follows:

1. MWGCN-1: This variant is configured to operate without the multi-scale wavelet decomposition module, which is essential for decomposing input data into different frequency components. Removing this module allows for the evaluation of the extent to which multi-scale analysis contributes to the model's detection capabilities.
2. MWGCN-2: In this variant, the scale-specific correlation learning module is excluded. This module typically improves model performance by learning complex dependencies between different data scales. Its removal helps to determine the impact of scale-specific learning on the overall accuracy and efficiency of the system.
3. MWGCN-3: By omitting the scale-attention module, this variant tests the importance of prioritizing specific scales over others and how this attention mechanism influences the model's ability to accurately identify icing conditions.
4. MWGCN-4: This configuration lacks the class rebalancing loss function, which is designed to manage imbalanced datasets by focusing more on minority classes. Removing this function allows for the assessment of its role in enhancing model fairness and effectiveness, particularly in scenarios where one class significantly outnumbers the other.

Table 6.4 presents the findings from the ablation study conducted on the two datasets, demonstrating that the complete MWGCN model outperforms all of its variants in terms of F1-score. This superior performance validates the integral role of each component in the MWGCN, contributing to the overall enhancement of model efficacy. The ablation results further indicate that the removal of any component from the MWGCN leads to a decrease in performance, underscoring the necessity of each module for optimal functionality.

**Table 6.4** Ablation analysis of the proposed MWGCN

| Dataset | Turbine-1 | | | Turbine-2 | | |
|---|---|---|---|---|---|---|
| | Precision | Recall | F1-score | Precision | Recall | F1-score |
| MWGCN-1 | 80 | 65 | 71 | 77 | 70 | 74 |
| MWGCN-2 | 80 | 67 | 73 | 67 | 55 | 60 |
| MWGCN-3 | 76 | 60 | 67 | 69 | 75 | 72 |
| MWGCN-4 | 80 | 62 | 70 | 78 | 59 | 67 |
| MWGCN | **76** | **74** | **75** | **88** | **72** | **79** |

Specifically, the performance differences across the variants highlight the unique contributions of each component to the model's effectiveness:

1. MWGCN-3 exhibits the most significant performance reduction on the Turbine-1 dataset, emphasizing the critical role of the scale-attention module in this context. The removal of this module impairs the model's capacity to appropriately weight the importance of different scales, which is particularly detrimental in scenarios like Turbine-1, where the input data displays considerable scale variation.
2. Conversely, MWGCN-2 shows the lowest performance on the Turbine-2 dataset, suggesting that the scale-specific correlation learning module is especially vital there. This indicates that inter-variable correlations at different scales are distinctly impactful for Turbine-2, possibly due to unique operational or environmental conditions specific to this turbine.

These variations in the impact of different components between datasets underscore the adaptive nature of the MWGCN, where the significance of each module may vary based on the specific characteristics and challenges presented by different datasets. The findings reinforce the importance of each proposed component in achieving high performance and suggest that the full integration of all modules is necessary for the MWGCN to function effectively across diverse turbine datasets.

### 6.3.6 Discussion

**(1) What are the main differences between the proposed MWGCN and the others?**

In this study, we have presented a blade icing detection model, the Multi-scale Wavelet Graph Convolutional Network (MWGCN), which utilizes deep learning techniques to exploit rich and implicit information across multiple scales in both the time and frequency domains. This model sets itself apart from traditional methodologies such as MSPCA, RF-XGBoost, LSTM, MLSTM-FCN, FCN, and SSENET, which primarily concentrate on modeling the temporal dynamics of original signals without investigating scale-based complexities. Unlike standard convolutional neural networks that are limited to a fixed receptive field determined by filter size, the MWGCN integrates a discrete wavelet decomposition module. This module enhances the model's capacity to capture a broader range of information across various time and frequency scales, significantly enriching the input for subsequent convolutional processing stages.

While comparable models like WaveletFCNN and MRCNN also employ discrete wavelet decomposition, they do not fully capitalize on inter-variable correlations within multi-scale SCADA data. The MWGCN addresses this limitation by incorporating a graph convolution module and a scale-attention module. These modules not only process structural and scale-specific information but also effectively model the complex interdependencies among different scales. Consequently, the MWGCN delivers superior performance by more effectively capturing and integrating signal information from multivariate time series across both time and frequency domains. This holistic approach to data analysis is the primary reason for the MWGCN's enhanced effectiveness and outperformance of other existing methodologies in blade icing detection.

**(2) How does the multi-scale wavelet transformation work?**

In this research, we utilize the scale-specific correlation learning module to efficiently capture scale-dependent inter-variable correlations within the input data. This approach enables the construction of adjacency matrices that reflect these correlations for each scale, allowing for a detailed analysis of how relationships between variables evolve across different scales. For illustrative purposes, we examine wind speed data and its correlations across various scales:

1. At scale 0, the most prominent correlations are observed with the temperatures of pitch motors 1, 2, and 3; the yaw position; and the switching temperature of pitch 3. This scale primarily reflects the most dominant temporal trends within the multivariate time series.
2. At scale 1, the primary correlations shift to the angles of pitch 1, 2, and 3, along with the temperatures of pitch motors 1 and 2. At scales 2 and 3, the correlations continue to center on the angles of pitch 1, 2, and 3, but also incorporate environmental and nacelle temperatures.

This pattern suggests that at the coarsest scale (scale 0), wind speed is strongly correlated with operational temperatures, which are directly related to turbine performance. As we progress to higher scales, the model begins to highlight pitch angles, indicating that these elements play a critical role in responding to more immediate, localized variations in wind conditions. This transition from temperature correlations at lower scales to angular correlations at higher scales demonstrates the model's ability to differentiate between stable, long-term trends and transient, dynamic changes within the dataset. The automatic learning of inter-variable relationships at multiple wavelet scales by the scale-specific correlation learning module not only enhances our understanding of the underlying dynamics within the data but also significantly improves classification performance by leveraging nuanced, scale-specific insights. This methodological innovation emphasizes the effectiveness of incorporating wavelet-based scale analysis in complex system monitoring, such as wind turbine operation.

**(3) How do hyperparameters influence performance?**

From Fig. 6.2, we can analyze how these hyperparameters affect model performance.

When evaluating the optimal Multi-scale Discrete Wavelet Decomposition (MDWD) scale for the proposed Multi-scale Wavelet Graph Convolutional Network (MWGCN), the most effective performance is attained at an MDWD scale of 3. Increasing the MDWD scale beyond this point tends to degrade the model's performance. This degradation may be attributed to the model's reduced ability to effectively process and utilize the information from higher level detail coefficients.

These coefficients, instead of serving as valuable inputs, may introduce additional complexity or noise that confounds the deep learning model, thereby diminishing its effectiveness.

Regarding the maximum number of neighbors in the scale-specific correlation learning module, the results confirm that a sparsification strategy—limiting the number of neighboring nodes each node considers—can enhance model performance. This approach not only reduces the computational demands of graph convolution processing but also minimizes the influence of noise, contributing to a more robust model. The observation that aggregating information from a select few nodes often yields better outcomes than including all nodes indicates that excessive node connections might introduce noise. Thus, a sparsification strategy can effectively mitigate such noise, reinforcing the model's robustness and accuracy.

Concerning the depth of the Graph Convolutional Network (GCN) layers, findings suggest that deeper GCN layers do not necessarily result in improved performance. Research by Li et al. (2018) and Xu et al. (2018) on graph representation learning reveals that, unlike Convolutional Neural Networks (CNNs), GCNs do not inherently benefit from increased depth [54, 55]. This limitation stems from the information aggregation mechanism in GCNs acting as a low-pass filter, as discussed by Nt. et al. (2019) [56]. With each additional layer, node signals become increasingly smoothed—a process that, while initially beneficial, can lead to over-smoothing when excessively repeated. This over-smoothing may cause the signals of individual nodes to converge, losing their unique characteristics and diminishing the diversity necessary for effective learning. Therefore, selecting an appropriate number of GCN layers is critical to maintaining the distinctiveness of node signals and ensuring optimal model performance.

## 6.4 Conclusion

Blade icing detection is essential for preserving the operational efficiency of wind turbines. In this paper, we introduce a wavelet-driven multi-scale graph convolutional network (MWGCN) specifically tailored for detecting icing conditions on wind turbines. To enhance the model's detection capabilities, we integrated discrete wavelet decomposition to capture multi-scale features of SCADA data across both time and frequency domains. Additionally, a scale-specific correlation learning module was implemented to infer inter-variable relationships at multiple scales, further deepening the model's analytical capacity. To improve processing efficiency without sacrificing accuracy, a temporal convolution submodule was incorporated for enhanced parallel processing capabilities. A scale-attention module was also introduced to facilitate the capture of inter-scale correlations, enriching the model's contextual understanding of the data. To address inherent class imbalances within the input data, the MWGCN includes a class rebalancing classifier, ensuring more equitable and precise model evaluations. Comprehensive experiments conducted with real-world datasets demonstrate the proposed model's effectiveness and its superiority over existing methods. Looking ahead, there are three promising directions for enhancing the MWGCN:

1. Cross-Scale Signal Correlation Exploitation: Further enhancements to the model could involve advanced techniques for exploiting cross-scale signal correlations, which would improve multivariate feature extraction capabilities.
2. Adaptive Hyperparameter Tuning: Considering the diverse nature of wind turbines and varying environmental conditions, adaptive and automatic hyperparameter tuning using methods such as reinforcement learning could optimize model performance across different scenarios.
3. Expansion to Unsupervised and Semi-Supervised Learning: Currently, the MWGCN operates within a supervised framework that relies on fully labeled datasets. Future research will explore modifications to adapt the model for unsupervised and semi-supervised settings, expanding its applicability and alleviating deployment constraints.

These directions represent significant opportunities to advance the state of wind turbine monitoring systems, ensuring that the MWGCN remains at the forefront of technological advancements in this domain.

## References


1. Gao L, Tao T, Liu Y, Hu H (2021) A field study of ice accretion and its effects on the power production of utility-scale wind turbines. Renew Energy 167:917–928
2. EWE Association, et al (2011) EU Energy policy to 2050. EWEA
3. Battisti L (2015) Wind turbines in cold climates: icing impacts and mitigation systems. Springer
4. Gent RW, Dart NP, Cansdale JT (2000) Aircraft icing. Philos Trans R Soc London. Ser A: Math Phys Eng Sci 358(1776):2873–2911
5. Chou J-S, Chiu C-K, Huang I-K, Chi K-N (2013) Failure analysis of wind turbine blade under critical wind loads. Eng Fail Anal 27:99–118
6. Dalili N, Edrisy A, Carriveau R (2009) A review of surface engineering issues critical to wind turbine performance. Renew Sustain Energy Rev 13(2):428–438
7. Jasinski WJ, Noe SC, Selig MS, Bragg MB (1998) Wind turbine performance under icing conditions
8. Antikainen P, Peuranen S (2000) Ice loads, case study. The proceedings of BOREAS V, Levi, Finland, vol 11
9. Tammelin B, Seifert H (2000) Wind energy production in cold climate
10. Wang P, Zhou W, Bao Y, Li H (2018) Ice monitoring of a full-scale wind turbine blade using ultrasonic guided waves under varying temperature conditions. Struct Control Health Monit 25(4):e2138
11. Muñoz CQG, Márquez FPG, Tomás JMS (2016) Ice detection using thermal infrared radiometry on wind turbine blades. Measurement 93:157–163
12. Muñoz CQG, Jiménez AA, Márquez FPG (2018) Wavelet transforms and pattern recognition on ultrasonic guides waves for frozen surface state diagnosis. Renew Energy 116:42–54
13. Tian W, Cheng X, Li G, Shi F, Chen S, Zhang H (2021) A multilevel convolutional recurrent neural network for blade icing detection of wind turbine. IEEE Sens J 21(18):20 311–20 323
14. Wei K, Yang Y, Zuo H, Zhong D (2020) A review on ice detection technology and ice elimination technology for wind turbine. Wind Energy 23(3):433–457

15. Roberge P, Lemay J, Ruel J, Bégin-Drolet A (2018) A new atmospheric icing detector based on thermally heated cylindrical probes for wind turbine applications. Cold Reg Sci Technol 148:131–141
16. Leite GdNP, da Cunha GTM, dos Santos Junior JG, Araújo AM, Rosas PAC, Stosic T, Stosic B, Rosso OA (2021) Alternative fault detection and diagnostic using information theory quantifiers based on vibration time-waveforms from condition monitoring systems: application to operational wind turbines. Renew Energy 164:1183–1194
17. Pérez JMP, Márquez FPG, Hernández DR (2016) Economic viability analysis for icing blades detection in wind turbines. J Cleaner Prod 135:1150–1160
18. Laakso T, Holttinen H, Ronsten G, Tallhaug L, Horbaty R, Baring-Gould I, Lacroix A, Peltola E, Tammelin B (2003) State-of-the-art of wind energy in cold climates. IEA Annex XIX 24:53
19. Parent O, Ilinca A (2011) Anti-icing and de-icing techniques for wind turbines: critical review. Cold Reg Sci Technol 65(1):88–96
20. Corradini ML, Cristofaro A, Pettinari S (2016) A model-based robust icing detection and estimation scheme for wind turbines. In: 2016 European control conference (ECC). IEEE, pp 1451–1456
21. Hu L, Zhu X, Chen J, Shen X, Du Z (2018) Numerical simulation of rime ice on nrel phase vi blade. J Wind Eng Ind Aerodyn 178:57–68
22. Davis NN, Byrkjedal Ø, Hahmann AN, Clausen N-E, Žagar M (2016) Ice detection on wind turbines using the observed power curve. Wind Energy 19(6):999–1010
23. Qiu Y, Feng Y, Sun J, Zhang W, Infield D (2016) Applying thermophysics for wind turbine drivetrain fault diagnosis using scada data. IET Renew Power Gener 10(5):661–668
24. Xu J, Tan W, Li T (2020) Predicting fan blade icing by using particle swarm optimization and support vector machine algorithm. Comput Electr Eng 87:106751
25. Tautz-Weinert J, Watson SJ (2017) Using scada data for wind turbine condition monitoring–a review. IET Renew Power Gener 11(4):382–394
26. Dai J, Yang W, Cao J, Liu D, Long X (2018) Ageing assessment of a wind turbine over time by interpreting wind farm scada data. Renew Energy 116:199–208
27. Dao PB, Staszewski WJ, Barszcz T, Uhl T (2018) Condition monitoring and fault detection in wind turbines based on cointegration analysis of scada data. Renew Energy 116:107–122
28. Song P, Yao Z, Zhao C (2019) Section division and multi-model method for early detection of icing on wind turbine blades. In: 2019 34rd youth academic annual conference of Chinese association of automation (YAC). IEEE, pp 749–754
29. Skrimpas GA, Kleani K, Mijatovic N, Sweeney CW, Jensen BB, Holboell J (2016) Detection of icing on wind turbine blades by means of vibration and power curve analysis. Wind Energy 19(10):1819–1832
30. Yi H, Jiang Q, Yan X, Wang B (2020) Imbalanced classification based on minority clustering smote with wind turbine fault detection application. IEEE Trans Ind Inform
31. Tao T, Liu Y, Qiao Y, Gao L, Lu J, Zhang C, Wang Y (2021) Wind turbine blade icing diagnosis using hybrid features and stacked-xgboost algorithm. Renew Energy 180:1004–1013
32. Xiao J, Li C, Liu B, Huang J, Xie L (2022) Prediction of wind turbine blade icing fault based on selective deep ensemble model. Knowl-Based Syst 242:108290
33. Cheng X, Li G, Ellefsen AL, Chen S, Hildre HP, Zhang H (2020) A novel densely connected convolutional neural network for sea-state estimation using ship motion data. IEEE Trans Instrum Meas 69(9):5984–5993
34. Jiang G, He H, Yan J, Xie P (2018) Multiscale convolutional neural networks for fault diagnosis of wind turbine gearbox. IEEE Trans Ind Electron 66(4):3196–3207
35. Chen L, Xu G, Zhang Q, Zhang X (2019) Learning deep representation of imbalanced scada data for fault detection of wind turbines. Measurement 139:370–379
36. Liu Y, Cheng H, Kong X, Wang Q, Cui H (2019) Intelligent wind turbine blade icing detection using supervisory control and data acquisition data and ensemble deep learning. Energy Sci Eng 7(6):2633–2645
37. Cheng X, Shi F, Zhao M, Li G, Zhang H, Chen S (2021) Temporal attention convolutional neural network for estimation of icing probability on wind turbine blades. IEEE Trans Ind Electron

38. Cheng X, Shi F, Liu X, Zhao M, Chen S (2021) A novel deep class-imbalanced semisupervised model for wind turbine blade icing detection. IEEE Trans Neural Netw Learn Syst
39. Yuan B, Wang C, Jiang F, Long M, Philip SY, Liu Y (2019) Waveletfcnn: a deep time series classification model for wind turbine blade icing detection, arXiv preprint arXiv:1902.05625v1
40. Percival DB (2008) Analysis of geophysical time series using discrete wavelet transforms: an overview. Nonlinear Time Series Anal Geosci 61–79
41. Nalley D, Adamowski J, Khalil B (2012) Using discrete wavelet transforms to analyze trends in streamflow and precipitation in quebec and ontario (1954–2008). J Hydrol 475:204–228
42. Nobre J, Neves RF (2019) Combining principal component analysis, discrete wavelet transform and xgboost to trade in the financial markets. Expert Syst Appl 125:181–194
43. Yuan B, Wang C, Luo C, Jiang F, Long M, Yu PS, Liu Y (2019) Waveletae: a wavelet-enhanced autoencoder for wind turbine blade icing detection, arXiv preprint arXiv:1902.05625
44. Chambers MJ (2016) The estimation of continuous time models with mixed frequency data. J Econom 193(2):390–404
45. Xu W, Liu W, Bian J, Yin J, Liu T-Y (2021) Instance-wise graph-based framework for multivariate time series forecasting, arXiv preprint arXiv:2109.06489
46. Mallat SG (2009) A theory for multiresolution signal decomposition: the wavelet representation. In: Fundamental papers in wavelet theory. Princeton University Press, pp 494–513
47. Wu Z, Pan S, Long G, Jiang J, Chang X, Zhang C (2020) Connecting the dots: multivariate time series forecasting with graph neural networks. In: Proceedings of the 26th ACM SIGKDD international conference on knowledge discovery & data mining, pp 753–763
48. Kingma DP, Ba J (2014) Adam: a method for stochastic optimization, arXiv preprint arXiv:1412.6980
49. Leahy K, Hu RL, Konstantakopoulos IC, Spanos CJ, Agogino AM, O'Sullivan DT (2018) Diagnosing and predictingwind turbine faults from scada data using support vector machines. Int J Prognostics Health Manag 9(1)
50. Sheriff MZ, Mansouri M, Karim MN, Nounou H, Nounou M (2017) Fault detection using multiscale pca-based moving window glrt. J Process Control 54:47–64
51. Zhang D, Qian L, Mao B, Huang C, Huang B, Si Y (2018) A data-driven design for fault detection of wind turbines using random forests and xgboost. IEEE Access 6:21 020–21 031
52. Karim F, Majumdar S, Darabi H, Harford S (2019) Multivariate lstm-fcns for time series classification. Neural Netw 116:237–245
53. Wang Z, Yan W, Oates T, Time series classification from scratch with deep neural networks: a strong baseline. In: 2017 international joint conference on neural networks (IJCNN). IEEE (2017), pp 1578–1585
54. Li Q, Han Z, Wu X-M (2018) Deeper insights into graph convolutional networks for semi-supervised learning. In Thirty-second AAAI conference on artificial intelligence
55. Xu K, Li C, Tian Y, Sonobe T, Kawarabayashi K-I, Jegelka S (2018) Representation learning on graphs with jumping knowledge networks. In: International conference on machine learning. PMLR, pp 5453–5462
56. Nt H, Maehara T (2019) Revisiting graph neural networks: all we have is low-pass filters. arXiv preprint arXiv:1905.09550

# Chapter 7
# Class-Imbalanced Graph-Temporal Convolutional Neural Network for Blade Icing Detection

## 7.1 Introduction

Wind energy, anticipated by the International Energy Agency to satisfy up to 50% of global electricity demand by the year 2050 [1], stands as a pivotal renewable energy resource. However, its widespread implementation, particularly in regions characterized by cold climates, is significantly hampered by the phenomenon of blade icing. This issue arises when supercooled water droplets solidify upon contact with rotating turbine blades, culminating in marked performance degradation and safety hazards, encompassing power reduction, aerodynamic instability, structural compromise, and ice shedding. Blade failures, the most frequently encountered turbine incidents, can precipitate power losses ranging from 20% to 50%, surpassing those attributed to other adverse conditions [2]. Blade icing also engenders turbine vibrations due to uneven ice shedding, thereby diminishing turbine longevity and curtailing power generation efficacy [3]. Scholarly works categorize icing impacts as complete shutdowns, aerodynamic disturbances, overloading, reduced fatigue lifespan, and safety risks [4]. A Swedish study (1998–2003) encompassing 1,337 turbine stoppages documented 161,523 hours of downtime, with blade icing responsible for 7% of incidents and an estimated 5% decrease in power output [5]. Long-term sequelae of icing include augmented turbine fatigue loads, jeopardizing structural robustness and operational lifespan [6]. Wind tunnel assessments reveal that even minimal ice accumulation can substantially disrupt blade aerodynamics [7], and atmospheric conditions can further intensify these ramifications [8]. Consequently, precise and timely blade icing detection is indispensable for effective and dependable wind turbine operation in frigid environments and for optimizing maintenance schedules. Addressing these challenges is vital for the sustained advancement and secure utilization of wind energy.

Contemporary advancements underscore the significance of accurate blade icing detection. Traditional methodologies rely on physical detection mechanisms such as ultrasonic damping, thermal infrared radiometry, and ultrasonic guided waves. While

 
X. Cheng et al., *Computational Methods for Blade Icing Detection of Wind Turbines*, Engineering Applications of Computational Methods 24,
https://doi.org/10.1007/978-981-96-6763-5_7

efficacious in capturing ice's physical characteristics and offering some degree of reliability, they present limitations. These approaches can potentially compromise blade integrity, are resource-intensive in terms of labor and materials, and often exhibit insensitivity to minor ice accumulations [9]. Consequently, indirect techniques, including power curve analysis, are employed to deduce icing and evaluate de-icing requirements. However, indirect methods are hindered by unstable internal sensors, leading to imprecise outcomes, and struggle to detect subtle ice levels, thereby diminishing their effectiveness for consistent detection. This emphasizes the pressing need for more efficient and precise detection methodologies in wind energy [10].

The constraints of conventional detection methods have spurred the evolution of signal monitoring techniques for blade icing detection, valued for their cost-effectiveness and minimal structural intrusion. These techniques are broadly classified as model-driven and data-driven strategies. Model-driven methods employ mathematical models to define the relationship between turbine operational signals, like rotor speed and ice accumulation [11]. These models, frequently expert-derived, furnish a theoretical framework for detection. Nevertheless, model-driven approaches possess limitations. Their applicability is often scenario-specific and heavily reliant on specialized expertise. Model refinement is costly, necessitating specialized equipment such as wind tunnels, thereby restricting practicality. Accuracy can also diminish under fluctuating environmental conditions, highlighting the demand for more adaptable and universally applicable ice detection methods.

Recent investigations into blade icing detection are increasingly oriented towards data-driven machine learning, particularly shallow and deep learning, propelled by the vast historical SCADA data. SCADA systems, primarily for wind turbine fault diagnosis, monitor operational parameters and signals, alongside environmental factors like temperature, wind speed, and direction [12]. This exhaustive dataset provides valuable insights into turbine conditions. Utilizing SCADA data for failure detection is pragmatic and economical, negating the need for additional sensors [10]. Deep learning methodologies are gaining prominence owing to their capacity to extract features and manage complex data structures [13]. Diverse strategies and models have been developed for the unique challenges of blade icing detection. For instance, Jiang et al. utilized a multi-scale learning CNN for broader turbine fault detection [14]. Chen et al. addressed class imbalance in icing detection by refining the learning process [15]. Liu et al. enhanced algorithm robustness by amalgamating ensemble and deep learning [16]. Cheng et al. advanced the domain with a temporal attention CNN and a semi-supervised model for blade icing detection [10, 17]. Yuan et al. developed WaveletFCNN, employing multi-scale wavelet analysis for feature extraction [18]. Tian et al. proposed MCRNN, a hybrid LSTM-CNN network capturing spatial and temporal data [9]. Lai et al. emphasized sensor correlation analysis via an adaptive graph learning module, noting that extant methods often overlook scale-specific inter-variable correlations across multiple wavelet scales [19]. Shallow learning techniques necessitate manual feature extraction, which is labor-intensive and costly. Deep learning automates much of this, augmenting efficiency and reducing resource needs, attracting substantial research interest. However, machine learning approaches exhibit considerable methodological variations.

Despite these deep learning advancements, critical challenges persist. Complex and dynamically evolving correlations among operational signals are not comprehensively understood. The pronounced class imbalance between "icing" and "non-icing" scenarios mandates specialized strategies for accurate and dependable detection. A wide array of weather conditions contribute to blade icing, with the frequency, duration, severity, and intensity of icing events significantly influencing generated signals. Detecting and utilizing anomalous changes in these patterns is essential for precise detection. Furthermore, the efficacy of blade icing detection via classification tasks is often undermined by training data imbalances, leading to biased outcomes. Addressing these challenges is paramount for enhancing deep learning algorithm accuracy and reliability in blade icing detection.

In this study, we introduce a Spatial-Temporal Graph Convolutional Network (SGCN) model specifically tailored for wind turbine blade icing detection, addressing the aforementioned challenges. The model integrates an innovative graph learning module employing self-attention to dynamically capture latent correlations among multiple signals, generating crucial graph information for the GCN component. To mitigate data imbalance, we propose a distance-based classifier designed for skewed data distributions, enhancing model effectiveness in diverse and complex conditions.

The contributions of this study are

- We introduce a Spatial-Temporal Graph Convolutional Network (SGCN) that autonomously learns dynamic correlations among multiple signals via a graph learning module founded on self-attention. This model captures both spatial and temporal dependencies between adjacent nodes through a spatial-temporal graph convolutional framework. Additionally, we propose a distance-based classifier formulated to address data imbalance by assigning distinct weights to each class based on their distances from the decision boundary.
- We perform extensive experiments utilizing real-world SCADA data from wind farms to validate the proposed model. The results substantiate the model's effectiveness, highlighting its superior performance and enhanced capability compared to existing methods.

The subsequent sections of this paper are structured as follows: Sect. 7.2 presents a detailed exposition of the proposed model. Section 7.3 deliberates the experimental findings, including evaluations of the model on public datasets and comparative analyses with baseline models on blade icing datasets. Finally, Sect. 7.4 concludes the paper and suggests potential directions for future research.

## 7.2 Methodology

### 7.2.1 Overview

In this study, we present a novel methodology for blade icing detection, encompassing four principal modules: the data preprocessing module, the graph learning module, the residual temporal graph module, and the icing detection module.

The initial dataset frequently contains considerable noise and missing values, rendering data cleansing an indispensable preliminary phase to mitigate these issues. Another pivotal objective of data preprocessing is to ascertain the data distribution and pinpoint variables that exhibit strong correlations with blade icing. The concluding stage of this module is data labeling, typically executed by domain experts.

The graph learning module processes multivariate time-series data utilizing a self-attention-based graph neural network that operates independently of a predefined graph structure. Initially, the time-series data are channeled through a Gated Recurrent Unit (GRU) layer, which computes hidden states at each time step, with the final hidden state representing the entirety of the time series. This strategy integrates a self-attention mechanism to capture latent interdependencies within the data, which constitutes the basis for constructing the adjacency weight matrix of the graph.

Our methodology incorporates a Spatial-Temporal Graph Convolutional Model (STGCM) to capture both spatial and temporal dependencies among adjacent nodes. The STGCM comprises a graph convolution layer, two gated convolution layers, a residual structure, and a batch normalization layer.

In the concluding phase, a distance-based classification approach is employed to generate logit vectors for each class. Subsequently, a loss function is utilized to quantify the disparity between the predicted values and the actual labels. To overcome the limitations of conventional loss functions, such as softmax loss, in addressing class imbalances, we propose a weighted loss function. This function is specifically designed to alleviate the impact of class imbalance, thereby enhancing the overall performance of the classifier.

### 7.2.2 Graph Learning Module

Through the innovative fusion of Gated Recurrent Units (GRUs) and a self-attention mechanism within a Graph Neural Network (GNN) architecture, this advanced graph learning module (GLM) provides a sophisticated solution for analyzing multivariate time-series data, obviating the necessity for a predefined graph structure. The process commences with GRUs, which systematically process the time-series input $\in \mathbb{R}^{N \times T}$, capturing temporal dynamics and preserving crucial sequential information through the iterative computation of hidden states. The terminal hidden state $\mathbf{R}$ functions as a comprehensive representation of the time series, encapsulating its salient characteristics and temporal patterns. Subsequently, the self-attention mechanism dynamically constructs an adjacency weight matrix **W** by discerning latent correlations among features through linear projections of **R** and attention score computations. The adjacency matrix is defined as $W = \text{Softmax}\left(\frac{QK^T}{\sqrt{d}}\right)$, where $Q$ and $K$ represent the query and key matrices, respectively. This methodology uncovers an inherent graph structure within the data, aligning features based on their implicit interrelationships. The GNN then leverages this dynamically constructed graph, applying graph convolution operations to progressively refine feature representations by embedding contextual relationships. This integrated approach not only captures temporal evolution via the GRUs but also exhaustively explores hidden inter-feature dependencies through self-attention, enabling the GNN to attain deeper insights and augmented analytical precision.

By integrating these constituents, this methodology establishes a novel benchmark in time-series analysis, with extensive applicability in forecasting, anomaly detection, and beyond. It uniquely synthesizes both the sequential progression of data and the complex interdependencies between features, setting a standard for high-performance solutions in intricate multivariate time-series tasks.

### 7.2.3 Residual Temporal Graph Module

In the context of detecting blade icing on wind turbines, sensor data is pivotal for constructing a spatial-temporal map that faithfully depicts icing conditions. However, an exclusive emphasis on either temporal or spatial dependencies can yield incomplete or inconsistent representations. To circumvent this limitation, we introduce the Spatial-Temporal Graph Convolutional Module (STGCM), which integrates both spatial and temporal information for a more holistic analysis. The STGCM comprises several essential components: a graph convolution layer, two gated temporal convolution layers, a residual structure, and a batch normalization layer, as depicted in Fig. 7.1.

The graph convolution layer is expressly designed to capture spatial dependencies by aggregating information from neighboring nodes, thereby establishing a coherent spatial context throughout the graph. Concurrently, the stacked gated temporal convolution layers sequentially update node features by incorporating data from successive time slices, thus enhancing the model's capacity to detect evolving patterns over time. This integration of spatial and temporal feature extraction equips the STGCM with a more comprehensive understanding of icing conditions, resulting in improved consistency and accuracy in detection.

The gated convolutional layer in our model is mathematically formulated as

$$y_{gated} = (K_1 * x_{gated} + b_1) \odot \sigma(K_2 * x_{gated} + b_2), \tag{7.1}$$

where $K_1$ and $K_2$ are convolutional kernels, and $b_1$ and $b_2$ represent their corresponding biases. The sigmoid function $\sigma$ is applied to modulate the input, thereby

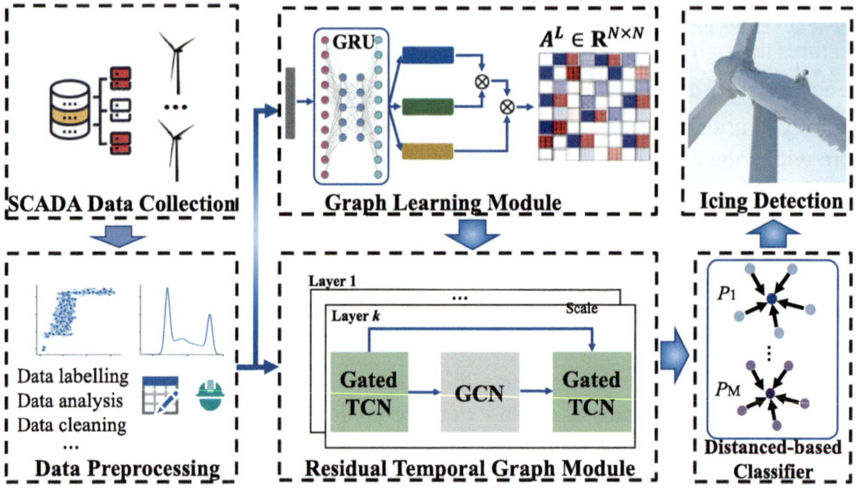


**Fig. 7.1** Overview structure of our proposed method

implementing a gating mechanism. Here, $x_{gated}$ denotes the input to the gated convolutional layer, while $y_{gated}$ denotes the output. It is important to note that $K_1$ and $K_2$ share identical dimensions, ensuring consistent processing across the channels.

Within our STGCM, the gated convolutional layers utilize this gating mechanism to selectively propagate information, thereby enhancing the extraction of temporal features. The incorporation of a residual structure further promotes efficient information flow by establishing direct connections between the input and output states. This design supports gradient preservation, mitigating the issue of vanishing gradients commonly encountered in deep architectures. Additionally, the batch normalization layer standardizes the feature distributions to a consistent scale, effectively reducing internal covariate shift. This normalization process contributes to the model's stability, accelerates convergence, and ultimately improves the overall performance of the STGCM.

### *7.2.4 Distance-Based Classifier*

The standard training paradigm entails employing a deep neural network to transform raw data into a feature representation. This representation, denoted as $h(x) \in \mathbb{R}^{N\times F}$, encapsulates the essential characteristics of the input data. Subsequently, a classifier comprising a single fully connected layer generates a logit vector for each class, where the classifier parameters are represented as $W \in \mathbb{R}^{F\times k}$. The operation of the classifier can be mathematically expressed as

$$l = W^T h(x) = [w_1^T h(x); w_2^T h(x); \cdots ; w_k^T h(x)], \tag{7.2}$$

where $w_i$ denotes the weight vector associated with the $i$-th class and $\mathcal{L}$ represents the logit vector for classification. To quantify the disparity between the predicted values and the ground-truth labels, we apply a softmax loss function across all classes. The softmax loss function is formulated as

$$\mathcal{L} = \sum_{i=1}^{k} \frac{N_i}{N} \mathcal{L}(D_i), \tag{7.3}$$

where $N_i$ represents the number of training samples in the $i$-th class, $N$ is the total number of samples, and $D_i$ denotes the set of samples belonging to the $i$-th class. This formulation adjusts the loss by accounting for class imbalances, weighing each class by its relative frequency in the dataset, thereby fostering balanced learning outcomes.

The softmax loss function, as defined in Eq. (7.3), is intrinsically susceptible to class imbalance, frequently resulting in a biased decision boundary that disproportionately favors the majority class. Specifically, in scenarios where the number of samples across different classes satisfies $N_1 \leq N_2 \leq \cdots \leq N_k$, it can be inferred that $\mathcal{L}_1 \leq \mathcal{L}_2 \leq \cdots \leq \mathcal{L}_k$. This relationship signifies that the classifier exhibits reduced sensitivity to misclassifications of minority class samples, leading to the establishment of a decision boundary that more accurately classifies majority class samples, often at the detriment of minority class accuracy.

To overcome the limitations of the conventional softmax loss function, we propose an alternative methodology based on a distance-based classifier (DBC). The DBC functions by mapping samples to computed class centroids based on their respective distances, facilitating a more balanced classification process. Various methods can be employed to calculate the likelihood of a sample belonging to a given class, including similarity metrics such as Euclidean distance and Cosine similarity, which offer versatile means for assessing similarity. From a technical standpoint, the feature learned by the residual temporal graph module is initially averaged and subsequently passed through a Multi-Layer Perceptron (MLP) with two linear layers to compute class centers. This procedure can be mathematically represented as follows:

$$\mathbf{c}_i = \frac{1}{n^i} \sum_{k=1}^{n^i} \Omega(X^{i,k}). \tag{7.4}$$

Let $\mathbf{c}_i \in R^{1 \times F_i}$ denote the center of the $i$-th class, where $i \in [1, \ldots, n_c]$ represents the total number of classes, and $n^i$ denotes the number of learned features in the latent space for the $i$-th class. Additionally, $X^i$ represents the feature set corresponding to the $i$-th class in the latent space. The parameters of the Multi-Layer Perceptron (MLP), denoted as $\Omega$, facilitate the calculation of the class centers.

Let $H_s \in \mathbb{R}^{F \times L}$ represent the feature representation of a sample, where $F$ is the dimensionality of the feature space and $L$ is the number of features. The probability that this sample belongs to class $i$ is computed as follows:

$$p(y = i|\mathcal{H}_s) = \frac{exp(- \parallel \mathcal{H}_s - \mathbf{c}_i \parallel^2)}{\sum_k exp(- \parallel \mathcal{H}_s - \mathbf{c}_i \parallel^2)}, \tag{7.5}$$

where $\parallel \cdot \parallel$ denotes the chosen distance function between $\mathcal{H}_s$ and $\mathbf{c}_i$.

### 7.2.5 Theoretical Analysis

In this section, we conduct a theoretical analysis to demonstrate the effectiveness of our model in handling class-imbalance learning. Let $\mathcal{D}$ represent a dataset consisting of positive (minority) samples $\mathcal{P}$ and negative (majority) samples $\mathcal{N}$, such that $\mathcal{P} \bigcup \mathcal{N} = \mathcal{D}$ and $\mathcal{P} \bigcap \mathcal{N} = \emptyset$. The number of positive and negative instances is $| \mathcal{P} |= N_p$ and $| \mathcal{N} |= N_n$, respectively, with $N_n \gg N_p$ due to class imbalance.

In this section, we furnish a theoretical analysis to demonstrate the efficacy of our model in addressing the challenges associated with class-imbalance learning. Let $\mathcal{D}$ denote a dataset comprising positive (minority) samples $\mathcal{P}$ and negative (majority) samples $\mathcal{N}$, such that $\mathcal{P} \bigcup \mathcal{N} = \mathcal{D}$ and $\mathcal{P} \bigcap \mathcal{N} = \emptyset$. The count of positive and negative instances is given by $| \mathcal{P} |= N_p$ and $| \mathcal{N} |= N_n$, respectively, with $N_n \gg N_p$, reflecting the typical class-imbalance scenario.

For any positive sample $\forall x_i \in \mathcal{P}$, the loss function is defined as follows:

$$\begin{aligned}\mathcal{L}(x_i) &= -\log\left[\frac{\exp(-d(h_s, \mathbf{c}1))}{\sum j = 0, 1 \exp(-d(h_s, \mathbf{c}j))}\right], \\ &= \log\left[1 + \exp(-d(h_s, \mathbf{c}0) + d(h_s, \mathbf{c}_1))\right]\end{aligned} \tag{7.6}$$

where $d(u, v) =\parallel u - v \parallel^2$ represents the squared Euclidean distance and $h_s$ is the output of the model for the given input sample.

Therefore, the cumulative loss for all positive samples in $\mathcal{P}$ is

$$\mathcal{L}(\mathcal{P}) = \sum_{i=1}^{N_p} \mathcal{L}(x_i) = \sum_{i=1}^{N_p} \log\left[1 + \exp(-d(h_i, \mathbf{c}0) + d(h_i, \mathbf{c}1))\right] \ . \tag{7.7}$$

Since $log(1 + exp(t))$ is a convex function with respect to $t$, we can apply Jensen's inequality to further simplify and bound the loss:

$$\begin{aligned}&= N_p log[1 + exp(-\frac{1}{N_p}\sum_{i=1}^{N_p} < h_i - \mathbf{c}_0, h_i - \mathbf{c}_0 > + \frac{1}{N_p}\sum_{i=1}^{N_p} < h_i - \mathbf{c}_1, h_i - \mathbf{c}_1 >)] \\ &= N_p log[1 + exp(-\mathbf{c}_0^T\mathbf{c}_0 + \mathbf{c}_1^T\mathbf{c}_1 + (\frac{2}{N_p}\sum_{i=1}^{N_p}(h_i^T(\mathbf{c}_0 - \mathbf{c}_1))))] \\ &= N_p log[1 + exp(- \parallel c_0 \parallel^2 + \parallel c_1 \parallel^2 + 2(\overline{h} - c_0)^T(c_0 - c_1)))].\end{aligned} \tag{7.8}$$

This equation holds if and only if, for all $i \neq j$, the following condition is satisfied: $-d(h_i, \mathbf{c}_0) + (h_i, \mathbf{c}_1) = -d(h_j, \mathbf{c}_0) + d(h_j, \mathbf{c}_1)$, where $< \cdot >$ denotes the inner product. Equation (7.8) attains its minimum value when the center $\mathbf{c}_1$ is defined as the mean of the positive samples, i.e.: $\mathbf{c}_1 = \frac{1}{N_p}\sum_{i=1}^{N_p} h_i$. In a similar manner, the loss for the negative samples can be computed. Thus, the total loss for the dataset $D\mathcal{D}D$ is expressed as

$$\begin{aligned}\mathcal{L}(\mathcal{N}) &= \sum_{i=1}^{N_n} \mathcal{L}(x_i) = \sum_{i=1}^{N_n} log[1 + exp(-d(h_i, \mathbf{c}_0) + d(h_i, \mathbf{c}_1))] \\ &\geq N_n log[1 + exp(-d(\mathbf{c}_1, \mathbf{c}_0))].\end{aligned} \tag{7.9}$$

Thus, the total loss is

$$\mathcal{L} = \mathcal{L}(\mathcal{N}) + \mathcal{L}(\mathcal{P}) \geq (N_p + N_n) log[1 + exp(-d(\mathbf{c}_1, \mathbf{c}_0))]. \tag{7.10}$$

Equation (7.10) defines the total loss function, $\mathcal{L}(\mathcal{P})$, for the proposed model. This total loss is the sum of the individual losses, $\mathcal{L}(x_i)$, for each sample in the positive class, where $N_p$ represents the total number of positive samples. Each individual loss $\mathcal{L}(x_i)$ is computed as the logarithm of one plus the exponential of the difference between two distances: the distance from the feature vector $h_i$ to the class center $\mathbf{c}0$ and the distance from the same feature vector to an alternative class center $\mathbf{c}1$.

The proposed model aims to minimize the total loss by reducing the individual losses, thereby drawing intra-class features closer to their respective class centers while simultaneously maximizing the distance between inter-class centers. Importantly, the change in total loss is governed primarily by the degree of intra-class feature clustering and the inter-class center distance, rather than by the sample size of each class. This property renders the model particularly effective in addressing the challenges associated with class imbalance.

### *7.2.6 Training Algorithm*

The Adam optimizer is employed for optimization, with a learning rate and weight decay set to 1e-4. In the configuration of the gated temporal convolutional network, the number of filters is set to 8. The model architecture comprises three layers, with each class being assigned a unique center. The training process spans 500 epochs.

**Algorithm 3** Training strategy

**Input:** SCADA data $X \in \mathbb{R}^{N \times T}$, the number of class $C$, number of layers $K$, number of prototypes $M$

**Output:** Blade icing prediction $\hat{y}$

Preprocess $X$ Initialize parameters of GLM, STGCM, and DBC modules **repeat**

$A \leftarrow GLM(X)$ **for** $i = 0;\ i < K;\ i = i + 1$ **do**

| $X \leftarrow STGCM_i(X, A, i)$

**end**

$X \leftarrow GlobalAveragePooling(X)$ **for** $i = 0;\ i < C;\ i = i + 1$ **do**

| Calculate $\mathbf{c}_i$ by using Eq. (7.4)

**end**

$p_0, p_1 \leftarrow DBC(X, \mathbf{c})$ Compute loss using Eq. (7.5) Update parameters using gradient descent

**until** *convergence*

$\hat{y} \leftarrow \arg\max p_0, p_1$

## 7.3 Experiment

The model is implemented utilizing the deep learning framework PyTorch, version 2.1.0. All experiments are performed on a server equipped with an NVIDIA Tesla V100 GPU.

### 7.3.1 *Experimental Settings*

#### 7.3.1.1 Dataset

To assess the performance of our proposed model, we employ two categories of datasets:

- **(1) Public Datasets**: We utilize 10 publicly accessible time-series classification datasets sourced from the UEA repository. The specifications of these datasets are detailed in Table 7.1.
- **(2) Blade Icing Dataset**: The blade icing dataset, provided by GOLDWIND, was compiled from two wind turbines situated in Inner Mongolia, China. The aggregate logging durations for these turbines were 305.77 hours and 695.59 hours, respectively. Domain specialists selected 26 key parameters for analysis, with comprehensive details on these parameters available in [10]. The labeling procedure was conducted by domain experts from GOLDWIND. For training and testing purposes, the data from both turbines were consolidated and partitioned into an 80% training and 20% testing split. Furthermore, three distinct imbalance ratios (IR) were established for this dataset to further evaluate the model's robustness.

**Table 7.1** Specification of 10 public time-series datasets

| Dataset | Abbr. | #Train | #Test | #Classes | Domain |
|---|---|---|---|---|---|
| ArticularyWord | AW | 275 | 300 | 25 | MOTION |
| AtrialFibrillation | AF | 15 | 15 | 3 | ECG |
| BasicMotions | BM | 40 | 40 | 4 | HAR |
| CharacterTrajectories | CT | 1422 | 1436 | 3 | MOTION |
| FaceDetection | FD | 5890 | 3524 | 2 | EEG/MEG |
| HandMovement | HM | 160 | 74 | 4 | EEG/MEG |
| Heartbeat | HB | 204 | 205 | 2 | AUDIO |
| NATOPS | NT | 180 | 180 | 6 | HAR |
| PEMS-SF | PS | 267 | 173 | 7 | PEMS |
| StandWalkJump | SWJ | 12 | 15 | 3 | HAR |

#### 7.3.1.2 Evaluation Metrics

For the public datasets, the widely adopted metric, **Accuracy**, is utilized to evaluate model performance. For the blade icing dataset, supplementary metrics are employed, including **Precision**, **Recall**, **F1-score**, and the **Matthews Correlation Coefficient (MCC)**. The definitions of these metrics are as follows:

$$Precision = \frac{TP}{TP + FP}, \quad Recall = \frac{TP}{TP + FN}, \tag{7.11}$$

$$F1 = \frac{2 \times Precision \times Recall}{Precision + Recall}, \tag{7.12}$$

$$MCC = \frac{TP \times TN - FP \times FN}{\sqrt{(TP + FP)(TP + FN)(TN + FP)(TN + FN)}}, \tag{7.13}$$

where $TP$, $FP$, $FN$, and $TN$ denote true positives, false positives, false negatives, and true negatives, respectively.

### 7.3.2 *Performance Comparison on Public Datasets*

We juxtapose the proposed model against six benchmark methods, elaborated as follows:

- **MLSTM-FCN (MF):** This technique categorizes time-series data by integrating Long Short-Term Memory (LSTM) networks with Fully Convolutional Networks (FCN), as detailed in [20].
- **EDI/DTWD/DTWI**: This approach extends the Dynamic Time Warping (DTW) algorithm, conventionally utilized for aligning time-series data, to accommodate

multidimensional datasets. The authors address the complexities inherent in generalizing DTW to higher-dimensional spaces and introduce several refinements to enhance its performance [21].

- **WEASEL+MUSE (WM)**: This technique classifies multivariate time-series data by combining the WEASEL (Word Embeddings for time-SEries cLassification) and MUSE (Multivariate Univariate Series Classification) algorithms, offering a robust methodology for capturing intricate patterns in time-series data [22].
- **TapNet (TPN)**: TapNet (Attentional Prototypical Network) is a deep learning-based method for classifying multivariate time-series data. By employing attention mechanisms, TapNet selectively emphasizes the most pertinent features within each time series, thereby improving classification accuracy [23].
- **CW-LSTM (CWL)**: This method introduces the Clockwork Recurrent Neural Network (CW-RNN), a variant of the Recurrent Neural Network (RNN) designed to address the challenge of learning long-term dependencies. In CW-RNN, the hidden state is updated at varied time intervals, enabling more effective capture of temporal dependencies [24].
- **TAMS-LSTM (TAMS)**: The Time-Aware Multi-Scale Recurrent Neural Network (TAMS-RNN) is an architecture that employs multiple RNNs operating at different time scales. This multi-scale approach enables TAMS-RNN to capture complex temporal dependencies in time-series data with enhanced accuracy [25].

Table 7.2 presents the outcomes of an experimental comparison between various baseline methods across 10 public datasets. The findings indicate that the proposed method attains the highest average performance, with a score of 0.799. Furthermore, it leads with the highest number of Wins/Ties, securing 7 wins/ties, outperforming all other methods. While the proposed method consistently delivers superior performance across the majority of datasets, it does not achieve the apex score on the AW, CT, and NT datasets, where it is surpassed by the WM, TPN, and TAMS methods, respectively.

Regarding performance on individual datasets, the proposed method achieves the highest score on two datasets: BM and FD, with scores of 1.000 and 0.676, respectively. However, it is pertinent to note that other methods, such as WM and TAMS, outperform the proposed method on certain other datasets. Overall, the results suggest that while the proposed method demonstrates robust performance across the datasets, its performance is not uniformly superior on every dataset, underscoring the necessity for a comprehensive evaluation that considers individual dataset results.

### 7.3.3 Comparison With Baseline Methods on Blade Icing Dataset

To evaluate the effectiveness of our proposed methodology for detecting icing on wind turbine blades, we conducted comparative analyses with five baseline models, described as follows:

**Table 7.2** Experimental results on public datasets

| Dataset | EDI | MF | DTWD | DTWI | WM | TPN | CWL | TAMS | Ours |
|---|---|---|---|---|---|---|---|---|---|
| AW | 97.0 | 97.3 | 98.7 | 98.0 | **99.0** | 98.7 | 94.7 | 97.3 | 97.1 |
| AF | 26.7 | 26.7 | 20.0 | 26.7 | 33.3 | 33.3 | 33.3 | 40.0 | **53.3** |
| BM | 67.5 | 95.0 | 97.5 | **100.0** | **100.0** | **100.0** | **100.0** | **100.0** | **100.0** |
| CT | 96.4 | 98.5 | 99.0 | 96.9 | 99.0 | **99.7** | 99.0 | 99.4 | 97.8 |
| FD | 51.9 | 54.5 | 52.9 | 51.3 | 54.5 | 55.6 | 59.4 | 60.2 | **67.6** |
| HM | 27.9 | 36.5 | 23.1 | 30.6 | 36.5 | 37.8 | 47.3 | 47.3 | **64.9** |
| HB | 62.0 | 66.3 | 71.7 | 65.9 | 72.7 | 75.1 | 72.2 | 75.6 | **76.1** |
| NT | 86.0 | 88.9 | 88.3 | 85.0 | 87.0 | 93.9 | 95.0 | **95.6** | 95.0 |
| PS | 70.5 | 69.9 | 71.1 | 73.4 | N/A | 75.1 | 89.0 | 89.0 | **93.6** |
| SWJ | 20.0 | 6.7 | 20.0 | 33.3 | 33.3 | 40.0 | 33.3 | 40.0 | **53.3** |
| Average | 60.6 | 64.0 | 64.2 | 66.1 | 68.4 | 70.9 | 72.3 | 74.4 | **79.9** |
| Win/Ties | 0 | 0 | 0 | 1 | 2 | 2 | 1 | 2 | **7** |

- **FCN**: This model utilizes three fully connected convolutional layers for time-series classification and has demonstrated superior performance in this domain [26].
- **MLSTM-FCN (MF):** This method integrates Long Short-Term Memory (LSTM) networks with Fully Convolutional Networks (FCN) for time-series classification, as outlined in [20].
- **DenseNet**: This approach employs a densely connected convolutional neural network, specifically designed for time-series classification, and has been successfully applied to sea-state estimation using ship motion data [13].
- **TACNN**: This model introduces the Temporal Attention Convolutional Neural Network (TACNN) architecture, which incorporates attention mechanisms to better capture temporal dependencies in time-series data [10].
- **MWGCN**: This method proposes a deep learning-based approach for detecting blade icing on wind turbines by introducing the Multi-scale Wavelet-driven Graph Convolutional Network (MWGCN) architecture. MWGCN integrates the strengths of graph convolutional networks with multi-scale wavelet analysis to improve detection accuracy [19].

Table 7.3 presents a performance comparison of five baseline algorithms for blade icing detection, with results reported for each algorithm across three distinct imbalance ratios (IRs) of positive to negative samples: 20:1, 50:1, and 100:1.

The results of our study indicate that the proposed algorithm surpasses the established baseline models, including FCN, LSTM-FCN, DenseNet, TACNN, and MWGCN, across all four evaluated metrics. Notably, at IRs of 50:1 and 100:1, our algorithm achieves the highest scores for both the F1-score and the Matthews Correlation Coefficient (MCC). At an IR of 20:1, TACNN registers the highest F1-score, demonstrating robust performance under lower imbalance conditions. However, even in this scenario, our algorithm remains competitive, particularly in terms of MCC.

**Table 7.3** Performance comparison with baselines on SCADA data

| Metrics | | FCN | MLSTM_FCN | DenseNet | TACNN | MWGCN | Ours |
|---|---|---|---|---|---|---|---|
| 20:1 | P | **100.000** | 96.000 | 80.000 | 58.400 | 71.500 | 78.900 |
| | R | **100.000** | 5.900 | 1.300 | 95.700 | 62.200 | 65.600 |
| | F1 | 16.000 | 11.200 | 1.300 | **72.500** | 68.100 | 71.600 |
| | MCC | 21.300 | 16.400 | 5.500 | 37.200 | 42.400 | **48.800** |
| 50:1 | P | **100.000** | **100.000** | 96.400 | 80.400 | 76.200 | 64.100 |
| | R | 6.600 | 3.200 | 4.400 | 32.400 | 47.100 | **75.500** |
| | F1 | 12.300 | 6.200 | 8.300 | 46.200 | 58.200 | **69.500** |
| | MCC | 18.400 | 12.800 | 14.100 | 30.500 | 35.000 | **41.500** |
| 100:1 | P | **100.000** | **100.000** | **100.000** | 63.500 | 67.700 | 66.900 |
| | R | 6.500 | 4.800 | 3.700 | 29.300 | 31.800 | **55.000** |
| | F1 | 12.100 | 9.200 | 7.100 | 40.000 | 43.300 | **60.400** |
| | MCC | 18.300 | 15.700 | 13.700 | 21.900 | 19.600 | **36.700** |

Overall, this comparative analysis provides compelling evidence for the efficacy of our proposed algorithm in blade icing detection.

### *7.3.4 Comparison With Class-Imbalance Learning Methods on Blade Icing Dataset*

To ascertain the efficacy of our proposed method, we conducted a comparative analysis utilizing five commonly employed loss functions designed for imbalanced learning. These methods are described as follows:

- **Focal Loss**: This widely recognized loss function was developed to address the challenges of learning under class imbalance [27]. In our implementation, we assign weights to each class based on its statistical representation.
- **Class-Balanced Loss (CB)**: This variant of focal loss incorporates the concept of the effective number of samples, as introduced by Cui et al. [28]. Our implementation closely adheres to the procedures outlined in their study.
- **Weighted Cross-Entropy (WCE)**: This modified version of the cross-entropy loss function assigns differential weights to each class, specifically mitigating issues related to class imbalance.
- **Label-Distribution-Aware Margin Loss (LDAM)**: Designed to handle class-imbalance scenarios, this innovative loss function replaces the conventional cross-entropy loss during training, thereby improving the handling of label distribution [29].
- **Balanced Softmax (BaSoft)**: This refined, unbiased adaptation of softmax is particularly well suited for learning in the presence of imbalanced classes, effec-

**Table 7.4** Performance comparison with imbalanced learning methods on SCADA data

| Metrics | | Focal | CB | WCE | LDAM | BaSoft | Ours |
|---|---|---|---|---|---|---|---|
| 20:1 | P | 61.4 | 77.6 | 71.9 | 45.9 | 61.6 | **78.9** |
| | R | **90.2** | 59.3 | 66.2 | 44.5 | 52.3 | 65.6 |
| | F1 | **73.1** | 67.2 | 68.9 | 45.2 | 56.5 | 71.6 |
| | MCC | 41.2 | 44.5 | 45.5 | 36.0 | 25.6 | **48.8** |
| 50:1 | P | 61.5 | 63.5 | 62.6 | 46.0 | 61.2 | **64.1** |
| | R | **84.4** | 66.9 | 70.3 | 44.5 | 46.2 | 75.5 |
| | F1 | 71.2 | 65.2 | **66.2** | 45.2 | 52.6 | 69.5 |
| | MCC | 35.9 | 34.4 | 36.5 | 22.5 | 19.8 | **41.5** |
| 100:1 | P | 60.5 | 65.9 | 65.4 | 46.7 | 58.8 | **66.9** |
| | R | **75.1** | 61.6 | 61.4 | 45.8 | 44.5 | 55.0 |
| | F1 | **67.1** | 63.7 | 63.4 | 46.3 | 50.7 | 60.4 |
| | MCC | 32.2 | 35.3 | 34.3 | 15.5 | 21.0 | **36.7** |

tively managing disparities in label distribution between training and testing datasets [30].

Table 7.4 presents a comparative analysis of various approaches to addressing imbalanced class distributions. The results demonstrate that our proposed method consistently outperforms the baseline methods in terms of the Matthews Correlation Coefficient (MCC) across all three imbalance ratios. Additionally, it achieves the highest Precision (P) and F1-score for the 20:1 and 100:1 ratios. However, at the 50:1 ratio, the F1-score of our method is marginally lower than that of the Weighted Cross-Entropy (WCE) method.

A more detailed examination of the results reveals that at the 20:1 imbalance ratio, our method exhibits superior precision and F1-score, effectively identifying minority class samples while minimizing false positives. The high Matthews Correlation Coefficient (MCC) further underscores the robustness and reliability of our approach in this setting. At the 50:1 ratio, although our method maintains the highest MCC, its F1-score is slightly lower than that of the Weighted Cross-Entropy (WCE) method, suggesting that WCE may provide a more favorable balance between precision and recall in this scenario. At the extreme 100:1 ratio, our method not only achieves the highest MCC but also registers the highest precision and F1-score, demonstrating its effectiveness in managing severely imbalanced class distributions. Overall, these results highlight the capability of our method to effectively manage imbalanced class distributions in SCADA data, outperforming other baseline approaches in most instances.

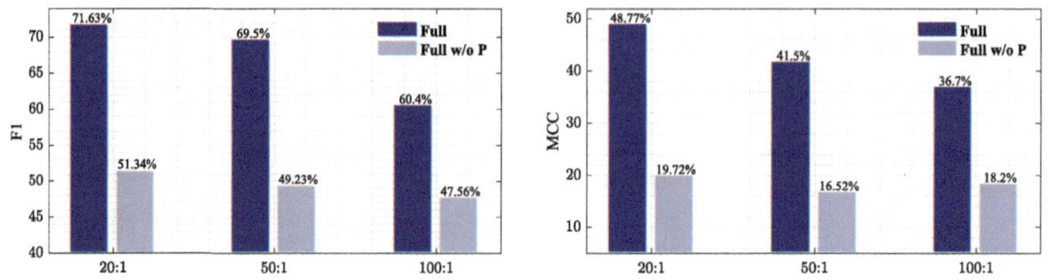


(a) F1 Score (b) Matthews Correlation Coefficient

**Fig. 7.2** Ablation study results

### *7.3.5 Ablation and Sensitivity Analysis*

To assess the significance of the proposed distance-based classifier, we created a variant of the model by substituting the distance-based classifier with a conventional softmax classifier, denoted as "Full w/o P". As depicted in Fig. 7.2, this modification results in a substantial performance decrement. Specifically, the absolute reductions in F1-score are 20.29%, 12.79%, and 9.19% across the three imbalance ratios (IRs). For the Matthews Correlation Coefficient (MCC), the respective decreases are 29.05%, 23.48%, and 17.78%.

Additionally, we conducted a sensitivity analysis to determine the optimal parameters for our model. In this analysis, we varied the number of layers and centers within the range of 1 to 3, with the results presented in Fig. 7.3. The analysis indicates that the model achieves peak performance in terms of both F1-score and MCC across all three imbalance ratios when configured with two layers and a single center.

### *7.3.6 Discussion*

This study introduces a novel approach for blade icing detection in wind turbines, leveraging class-imbalanced spatial-temporal feature learning. The proposed method comprises three key components: a graph learning module, a spatial-temporal graph convolutional network, and a distance-based classifier. The graph learning module autonomously constructs a graph to capture the correlations among different wind turbines. By integrating this graph structure with temporal data, the spatial-temporal graph convolutional network effectively extracts features crucial for blade icing detection. Subsequently, the distance-based classifier utilizes these features to differentiate between normal and abnormal samples, enabling accurate predictions. Comprehensive experiments were performed on both public datasets and a specialized blade icing dataset, providing a robust benchmark of our method against several state-of-the-art techniques. The results demonstrate the superior performance of our approach in terms of accuracy, precision, recall, F1-score, and Matthews Correlation

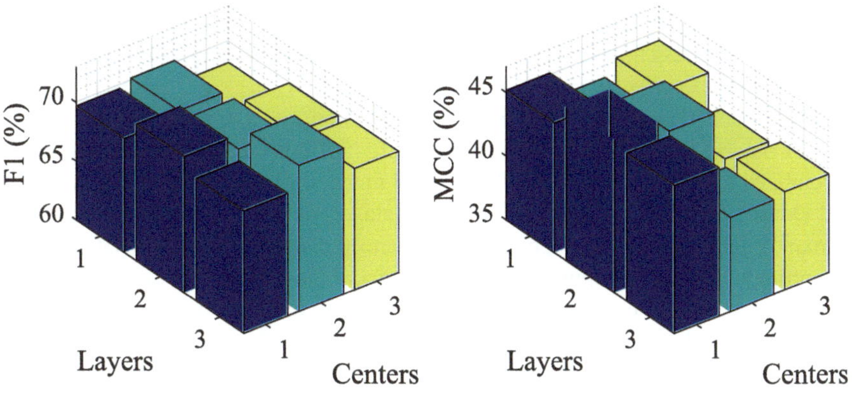


(a) Imbalance Ratio 20:1

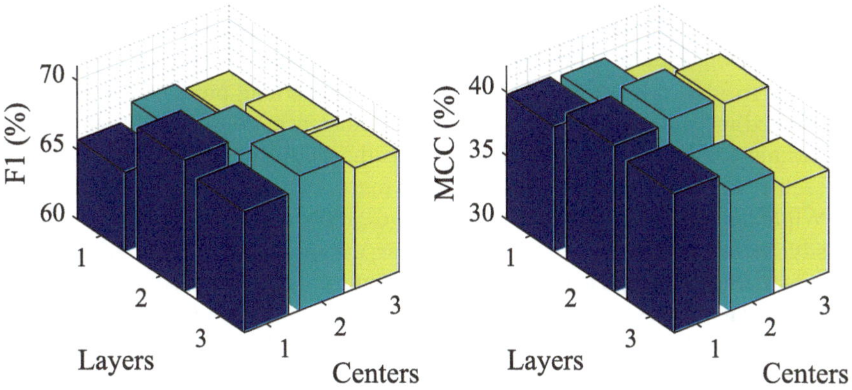


(b) Imbalance Ratio 50:1

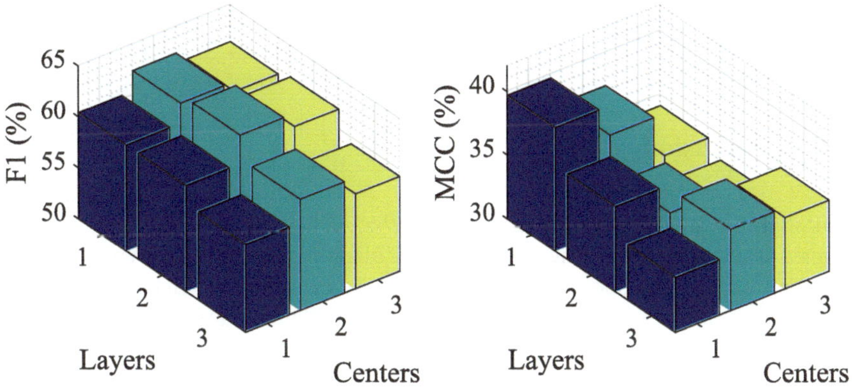


(c) Imbalance Ratio 100:1

**Fig. 7.3** Sensitivity analysis of model parameters across varying imbalance ratios

Coefficient (MCC). Furthermore, extensive ablation and sensitivity analyses were conducted to assess the contribution of each component to the overall performance of the methodology.

The enhanced performance of our model, in comparison to state-of-the-art methods for blade icing detection, can be principally attributed to the incorporation of a specialized imbalanced classifier. Traditional models, such as FCN, MLSTM-FCN, and DenseNet, are optimized for balanced datasets and experience significant performance degradation when applied to imbalanced datasets. While MWGCN shares commonalities with our approach, it employs focal loss to mitigate class imbalance, which necessitates meticulous parameter tuning—a challenge in optimizing performance for imbalanced learning tasks. In contrast, TACNN addresses imbalance by rebalancing the dataset prior to feature learning, limiting its direct applicability to datasets that are inherently imbalanced.

In contradistinction to these state-of-the-art loss functions, our method adopts a clustering-based approach to directly address the imbalanced features learned by the model, offering a more "direct" solution to handling data imbalance. In comparison, other approaches primarily focus on adjusting the learning process, which can be considered an "indirect" strategy for managing imbalance. Consequently, our method proves more effective, as it directly confronts the challenges posed by imbalanced datasets rather than relying on modifications to the learning process.

While our method offers several advantages, it also presents certain limitations. Firstly, we observed suboptimal convergence during the training process, which may arise from the graph learning mechanism introduced in this study. Additionally, experimental results suggest that the feature learning capability of the graph convolution module could be further optimized. Secondly, while the distance-based classifier effectively addresses imbalanced classification, its efficacy has primarily been validated in binary classification scenarios. The performance and applicability of this approach in multi-class imbalanced settings remain underexplored, highlighting an important area for future research and development.

## 7.4 Conclusion

This study substantiates the feasibility and effectiveness of graph learning methodologies in addressing the intricate challenge of blade icing detection in wind turbines. Furthermore, we introduce a novel distance-based classifier, which effectively alleviates the pervasive issue of class imbalance—a challenge often underestimated by existing methods. Through rigorous evaluation on both synthetic and real-world datasets, our approach establishes a new performance standard, offering valuable insights that could inform and inspire future research endeavors in this domain.

Several promising directions for future research emerge from our findings. One key avenue involves investigating more sophisticated graph learning strategies capable of capturing the temporal dynamics within wind turbine networks. Another direction is the integration of multiple data modalities, such as meteorological and visual data, to further enhance detection accuracy and robustness. Additionally, our approach could be adapted for broader industrial applications, particularly those requiring spatio-temporal analysis and anomaly detection.

## References

1. Cheng X, Shi F, Liu Y, Liu X, Huang L (2022) Wind turbine blade icing detection: a federated learning approach. Energy 254:124441
2. Chou J-S, Chiu C-K, Huang I-K, Chi K-N (2013) Failure analysis of wind turbine blade under critical wind loads. Eng Failure Anal 27:99–118
3. Alsabagh A, Tiu W, Xu Y, Virk MS (2013) A review of the effects of ice accretion on the structural behavior of wind turbines. Wind Eng 37(1):59–70
4. Dalili N, Edrisy A, Carriveau R (2009) A review of surface engineering issues critical to wind turbine performance. Renew Sustain Energy Rev 13(2):428–438
5. Hochart C, Fortin G, Perron J, Ilinca A (2008) Wind turbine performance under icing conditions. Wind Energy Int J Progress Appl Wind Power Conv Technol 11(4):319–333
6. Frohboese P, Anders A (2007) Effects of icing on wind turbine fatigue loads. J Phys Conf Ser 75(1). IOP Publishing, p 012061
7. Li Y, Tagawa K, Feng F, Li Q, He Q (2014) A wind tunnel experimental study of icing on wind turbine blade airfoil. Energy Conv Manag 85:591–595
8. Lamraoui F, Fortin G, Benoit R, Perron J, Masson C (2014) Atmospheric icing impact on wind turbine production. Cold Regions Sci Technol 100:36–49
9. Tian W, Cheng X, Li G, Shi F, Chen S, Zhang H (2021) A multilevel convolutional recurrent neural network for blade icing detection of wind turbine. IEEE Sensors J 21(18):20 311–20 323
10. Cheng X, Shi F, Zhao M, Li G, Zhang H, Chen S (2021) Temporal attention convolutional neural network for estimation of icing probability on wind turbine blades. IEEE Trans Indus Electron
11. Hu L, Zhu X, Chen J, Shen X, Du Z (2018) Numerical simulation of rime ice on nrel phase vi blade. J Wind Eng Indus Aerodyn 178:57–68
12. Xiao J, Li C, Liu B, Huang J, Xie L (2022) Prediction of wind turbine blade icing fault based on selective deep ensemble model. Knowl Based Syst 242:108290
13. Cheng X, Li G, Ellefsen AL, Chen S, Hildre HP, Zhang H (2020) A novel densely connected convolutional neural network for sea-state estimation using ship motion data. IEEE Trans Instrumen Measur 69(9):5984–5993
14. Jiang G, He H, Yan J, Xie P (2018) Multiscale convolutional neural networks for fault diagnosis of wind turbine gearbox. IEEE Trans Indus Electron 66(4):3196–3207
15. Chen L, Xu G, Zhang Q, Zhang X (2019) Learning deep representation of imbalanced scada data for fault detection of wind turbines. Measurement 139:370–379
16. Liu Y, Cheng H, Kong X, Wang Q, Cui H (2019) Intelligent wind turbine blade icing detection using supervisory control and data acquisition data and ensemble deep learning. Energy Sci Eng 7(6):2633–2645
17. Cheng X, Shi F, Liu X, Zhao M, Chen S (2021) A novel deep class-imbalanced semisupervised model for wind turbine blade icing detection. IEEE Trans Neural Netw Learn Syst
18. Yuan B, Wang C, Jiang F, Long M, Philip SY, Liu Y (2019) Waveletfcnn: a deep time series classification model for wind turbine blade icing detection. arXiv preprint arXiv:1902.05625v1
19. Lai Z, Cheng X, Liu X, Huang L, Liu Y (2022) Multiscale wavelet-driven graph convolutional network for blade icing detection of wind turbines. EEE Sensors J
20. Karim F, Majumdar S, Darabi H, Harford S (2019) Multivariate lstm-fcns for time series classification. Neural Networks 116:237–245

21. Shokoohi-Yekta M, Wang J, Keogh E (2015) On the non-trivial generalization of dynamic time warping to the multi-dimensional case. In: Proceedings of the 2015 SIAM international conference on data mining. SIAM, pp 289–297
22. Schäfer P, Leser U (2017) Multivariate time series classification with weasel+ muse. arXiv preprint arXiv:1711.11343
23. Zhang X, Gao Y, Lin J, Lu C-T (2020) Tapnet: multivariate time series classification with attentional prototypical network. Proc AAAI Conf Artif Intell 34(04):6845–6852
24. Koutnik J, Greff K, Gomez F, Schmidhuber J (2014) A clockwork rnn. In: International conference on machine learning. PMLR, pp 1863–1871
25. Chen Z, Ma Q, Lin Z (2021) Time-aware multi-scale rnns for time series modeling. In: IJCAI, pp 2285–2291
26. Wang Z, Yan W, Oates T (2017) "Time series classification from scratch with deep neural networks: a strong baseline. In: International joint conference on neural networks (IJCNN). IEEE, pp 1578–1585
27. Lin T-Y, Goyal P, Girshick R, He K, Dollár P (2017) Focal loss for dense object detection. In: Proceedings of the IEEE international conference on computer vision, pp 2980–2988
28. Cui Y, Jia M, Lin T-Y, Song Y, Belongie S (2019) Class-balanced loss based on effective number of samples. In: Proceedings of the IEEE conference on computer vision and pattern recognition, pp 9268–9277
29. Cao K, Wei C, Gaidon A, Arechiga N, Ma T (2019) Learning imbalanced datasets with label-distribution-aware margin loss. In: Advances in neural information processing systems, pp 1567–1578
30. Ren J, Yu C, Ma X, Zhao H, Yi S et al (2020) Balanced meta-softmax for long-tailed visual recognition. Adv Neural Inform Process Syst 33:4175–4186

# Part III
# Federated Convolutional Neural Network-Based Methods

As highlighted in the preceding part introduction, Part III shifts focus to explore Federated Convolutional Neural Network (FCNN)-based methods for blade icing detection. This transition acknowledges the increasing need for collaborative and privacy-preserving approaches, particularly relevant in distributed wind farm settings. While Parts I and II demonstrated the efficacy of centralized deep learning techniques, Part III delves into the transformative potential of Federated Learning (FL) to enable collaborative model training across geographically dispersed wind farms without requiring direct data sharing. These innovative methods harness the power of federated learning principles in conjunction with convolutional neural networks to address critical challenges inherent in collaborative blade icing detection, such as data heterogeneity across different wind farms, class imbalance within individual datasets, and paramount data privacy concerns. By leveraging FCNN architectures within a federated framework, Part III demonstrates a remarkable ability to build robust and generalizable icing detection models while maintaining data locality and security. Each method presented in Part III tackles a unique challenge within the realm of federated learning for blade icing detection, and is purposefully designed to contribute significantly to enhancing model performance, improving data efficiency, and bolstering security within distributed learning environments.

The core objective of Part III is to provide a comprehensive understanding of how Federated Convolutional Neural Network methodologies can be effectively employed for blade icing detection in realistic, real-world wind farm scenarios. These FCNN-based methods are specifically engineered to enable collaborative model training across multiple clients (individual wind farms) while rigorously preserving data privacy and proactively addressing challenges intrinsically linked to federated learning, such as significant data heterogeneity and prevalent class imbalance. By employing advanced techniques within the federated learning framework, including heterogeneous model architectures, knowledge distillation strategies, and innovative blockchain integration, the models are able to achieve robust, data-efficient, and demonstrably secure blade icing detection capabilities across distributed environments. Furthermore, Part III introduces novel approaches to mitigate practical constraints often encountered in federated settings, such as limited communication bandwidth between wind farms and varying computational resources available at each location. For example, Chap. 8 introduces the Split Heterogeneous Federated Convolutional Neural Network (BiFL), a novel federated learning paradigm designed to address data heterogeneity and resource constraints. BiFL employs heterogeneous model architectures, utilizing simpler client-side models and a more complex server-side model to optimize resource distribution and communication efficiency, while enhancing data privacy through split learning and tackling class imbalance with a prototype-based approach. Building on data efficiency considerations, Chap. 9 presents a Data-Efficient Heterogeneous Federated Convolutional Neural Network (FedBIP). This innovative FL approach leverages feature selection to prioritize informative features and reduce data dimensionality, incorporates over-sampling for data balancing, and employs a weighted aggregation method with knowledge distillation to accelerate learning and improve model performance in data-scarce federated settings. Addressing crucial security and transparency concerns, Chap. 10 introduces a Blockchain-Empowered Clustered Federated Learning Model (BLADE). This groundbreaking architecture innovatively integrates blockchain technology into a clustered federated learning framework to decentralize model aggregation and update processes, leveraging blockchain's inherent security features to mitigate data privacy risks and enhance model trustworthiness, while also addressing class imbalance through a cluster-based imbalanced learning algorithm.

Each chapter included in Part III meticulously details a unique Federated Convolutional Neural Network approach, substantiated by rigorous experimental evaluations conducted on real-world Supervisory Control and Data Acquisition (SCADA) data collected directly from operational wind turbine systems. The experimental findings comprehensively presented throughout this part emphatically underscore the considerable potential and practical applicability of Federated Learning, specifically utilizing FCNNs, within the critical context of blade icing detection. This lays a robust foundation for the realization of more scalable, secure, and privacy-preserving icing detection methodologies, essential for the widespread adoption of collaborative AI in the renewable energy sector. This research significantly enriches the wider understanding of federated learning strategies tailored for deployment in distributed and data-sensitive environments, particularly within the rapidly evolving renewable energy landscape.

# Chapter 8
# Split Heterogeneous Federated Convolutional Neural Network for Blade Icing Detection

## 8.1 Introduction

The domain of wind energy technology has witnessed substantial advancements, significantly propelling the progress of wind turbine systems. Despite these strides, blade icing persists as a critical impediment to the operational efficiency and dependability of these systems, particularly in high-latitude zones. Under extreme conditions, blade icing can precipitate a decline of up to 30% in total annual power generation [1].

Conventionally, blade icing detection has been tackled using three primary methodologies: human observation, passive methodologies, and active strategies. Human observation, however, is inherently subjective, as it relies on the observer's proficiency. Passive techniques, employing materials like black coatings or anti-icing layers [2], demonstrate limited effectiveness, given the impracticality of solely depending on coatings for prevention. Conversely, active methods, utilizing external sensors or mechanical devices [3], necessitate supplementary energy and mechanical components, potentially posing a risk of turbine damage [1].

In contrast to these established methods, mathematical and data-driven approaches have garnered increasing interest. Model-driven paradigms, employing mathematical models to define relationships between weather parameters and blade icing based on physical principles, have been investigated [4, 5]. Two prevalent categories include the empirical statistical model (ESM) and the ice growth model (IGM). While the ESM capitalizes on long-term regional monitoring data, its precision is frequently constrained. The IGM, conceptualizing blade icing as a multifaceted phenomenon encompassing aerodynamics, hydrodynamics, and heat transfer, offers valuable insights but mandates considerable domain-specific knowledge and experimental resources like wind tunnels [6]. Mathematical models, predicting icing through numerical formulations [7], are similarly restricted by their reliance on specific assumptions, potential inaccuracies, the necessity for experimental validation, and substantial domain expertise for effective interpretation [6] (Fig. 8.1).



X. Cheng et al., *Computational Methods for Blade Icing Detection of Wind Turbines*,
Engineering Applications of Computational Methods 24,
https://doi.org/10.1007/978-981-96-6763-5_8

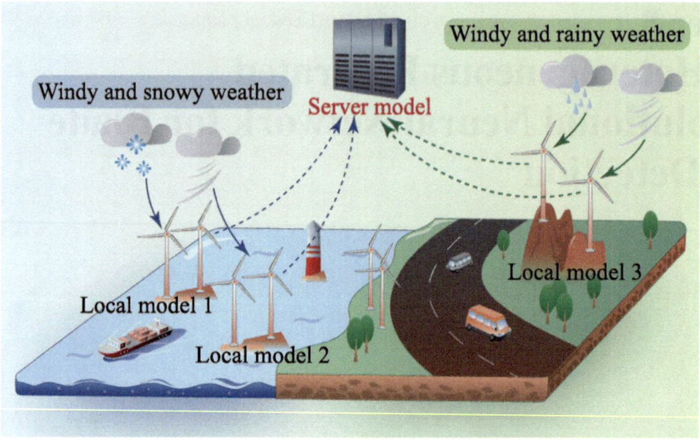


**Fig. 8.1** Illustration of the proposed BiFL for blade icing detection under different weather conditions

More recently, the extensive proliferation of Internet of Things (IoT) technologies has facilitated the acquisition of large-scale datasets, enabling the evolution of data-driven models for blade icing detection. Data-driven methodologies, notably deep learning techniques, have risen to prominence owing to their aptitude for autonomous feature extraction without prior domain knowledge. Illustrative examples include machine learning algorithms for ultrasonic signal classification [8], ensemble deep learning for SCADA data feature extraction [9], wavelet-based CNN models [10], and temporal CNN models [11]. A principal benefit of these models is their data-centric nature, often enhancing both efficiency and cost-effectiveness [11], leading to their escalated adoption in both academic and industrial spheres, particularly for end-to-end deep neural network solutions. However, conventional centralized implementations of these data-driven methods, where data from geographically dispersed wind farms are aggregated in central locations for model training, encounter challenges. These encompass substantial computational resource demands, extensive data storage requirements, considerable communication overhead, cybersecurity vulnerabilities, and reluctance from wind farms to share data due to competitive considerations [12].

Data-driven models, especially deep neural networks, necessitate voluminous datasets and significant computational resources for effective training. Integrating data from multiple wind farms can augment model performance by leveraging spatio-temporal interdependencies [12]. While centralized data consolidation has been the conventional approach, it suffers from limitations in computational capacity, network bandwidth, cybersecurity, and data sharing reluctance, culminating in the "data island" predicament. In response to escalating data protection concerns within the energy sector, Federated Learning (FL) has surfaced as a promising methodological alternative [13]. FL, as exemplified by frameworks for solar irradiation forecasting [14], distributed energy forecasting [15], sociodemographic prediction [16], and applications in European projects such as Smart4RES [17] and PHOENIX [18], facilitates the creation of global models by aggregating locally trained models, thereby circumventing raw data transfer. This research addresses the drawbacks of centralized approaches by introducing an FL-based framework for blade icing detection. This framework enables distributed model training across wind farms and eliminates the requirement for centralized data storage, representing, to the best of our knowledge, the inaugural application of FL to this specific problem.

Despite its merits, the application of traditional FL frameworks to blade icing detection encounters several constraints. Primarily, typical FL models mandate uniform neural network architectures across both server and clients. Wind turbines located in remote sites with limited computational resources present obstacles to training intricate models. Employing a small, shared network restricts learning potential, while deploying larger models becomes client-resource prohibitive. Consequently, a consistent architecture is impractical. Furthermore, the efficacy of icing detection models is sensitive to variations in local climatic conditions. While FL addresses regional diversity via aggregating local models, data heterogeneity and class imbalance introduce further complexities. Secondarily, sensor data from individual clients exhibit pronounced class imbalance, with non-icing scenarios dominating. This non-IID data distribution can adversely affect global model performance. Data heterogeneity, or non-IID data distribution, stems from regional weather disparities. Although solutions like FedProx [19], personalized FL [20, 21], and clustered FL [22, 23] address data heterogeneity, they do not entirely resolve the challenges in blade icing detection. Thirdly, blade icing is turbine dependent; smaller turbines might ice up whereas larger ones remain unaffected under comparable conditions. Conventional FL, premised on identical architectures, is inadequate in this context. Model heterogeneity, where clients utilize varying architectures, has been tackled through knowledge distillation (KD) [24, 25] and network structure search [26, 27]. However, these often overlook client computational limitations and the combined complexities of data and model heterogeneity.

To surmount these delineated challenges, we propose BiFL (Blade Icing Federated Learning), an innovative heterogeneous FL model for collaborative blade icing detection, tailored for diverse turbine types and weather scenarios. Diverging from traditional FL, BiFL employs distinct model architectures, allocating smaller networks to clients and a larger model to the server, thus optimizing resource distribution. BiFL transmits hidden features rather than gradients, diminishing communication overhead and bolstering privacy. To alleviate class imbalance, we present a prototype-based methodology, learning class prototypes from extracted features for classifier construction. BiFL also integrates transfer learning via knowledge distillation (KD) for efficient knowledge conveyance, enhancing robustness, generalization, and overall detection performance. To effectively manage both data and model heterogeneity, BiFL adopts a heterogeneous collaborative strategy, ensuring peak performance across resource-constrained clients and the server by utilizing smaller client-side models and a larger server-side model.

In essence, the contributions of this study are threefold:

1. We introduce BiFL, an FL-based approach for blade icing detection in wind turbines. To the best of our knowledge, this represents the pioneering application of FL within the domain of blade icing detection. BiFL leverages a heterogeneous architecture between clients and server to collaboratively train a global model, optimizing data utilization, guaranteeing privacy, and minimizing communication demands.
2. We put forth a novel prototype-based method to tackle class imbalance by equilibrating extracted features, in contrast to conventional data sampling or algorithmic modifications. Furthermore, we incorporate an information safeguarding mechanism to protect against potential data breaches without compromising model effectiveness.
3. We conducted a thorough evaluation of BiFL against state-of-the-art FL models and class-imbalance methodologies, demonstrating superior efficacy in blade icing detection. Ablation studies quantified the contributions of individual components, and sensitivity analyses scrutinized the impact of key hyperparameters.

The organization of this paper is structured as follows. Section 8.2 provides an exhaustive exposition of the proposed BiFL model. Section 8.3 delineates the evaluation methodology employed in this research. Finally, Sect. 8.4 concludes the paper and explores potential avenues for subsequent research.

## 8.2 Methodology

### *8.2.1 Overview*

Wind farms are frequently established in remote locales characterized by a scarcity of computational resources, such as GPUs or substantial memory capacities. Within a typical FL framework, the implementation of blade icing detection for wind turbines would necessitate considerable upgrades to the local computing infrastructure, resulting in elevated expenses. To circumvent this challenge, we introduce a heterogeneous FL model specifically adapted for blade icing detection, engineered to operate efficiently across diverse wind turbine types under varying meteorological conditions. Drawing inspiration from the framework presented in [28], we propose BiFL, a heterogeneous FL method designed to address class imbalance. As depicted in Fig. 8.2, BiFL encompasses $K$ local client models and a global server model. Each client, situated within a distinct climate zone, utilizes its locally sourced data to train a model. The insights derived from these local models are then communicated to a central server, where they are aggregated to construct a global model. This global model is subsequently disseminated back to the clients to refine their local models. Given the fluctuations in environmental conditions across wind turbine locations, the sensor data acquired by the SCADA system often exhibits high degrees of nonlinearity and non-stationarity. To mitigate these complexities, both client and

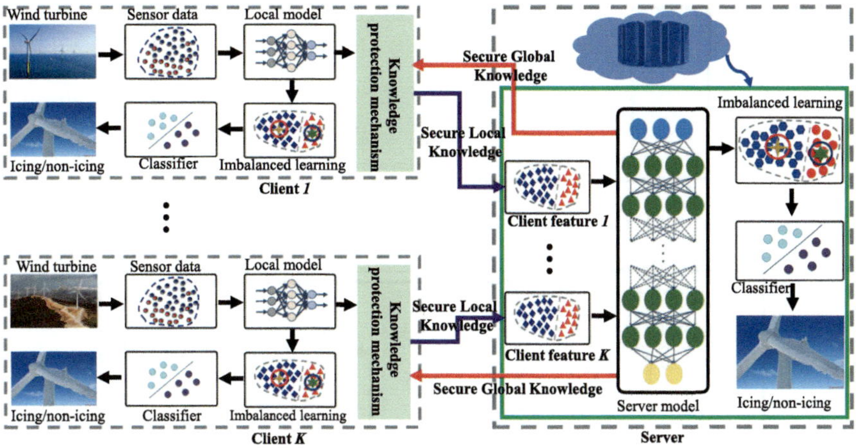


**Fig. 8.2** Structure of the proposed BiFL

server models employ Convolutional Neural Networks (CNNs) for efficacious feature extraction. Considering the constrained computational resources at the client end, the local models are intentionally designed to be significantly simpler than the global model deployed on the server. To optimize performance, we advocate for the utilization of a compact CNN for each client, while the server employs a more substantial, complex CNN better suited to its enhanced computational capabilities. This configuration facilitates periodic knowledge exchange between the local clients and the central server, ensuring that the global model capitalizes on the extensive data at the client level, while simultaneously accommodating the computational limitations inherent at the client side.

### 8.2.2 *Feature Extractor for Client and Server Model*

Given the disparity in computational prowess between clients and servers, it is imperative that their respective models exhibit variations in complexity. In this study, we propose employing a single-layer Convolutional Neural Network (SLCNN) for client models, whereas the server model utilizes a purpose-built densely connected Convolutional Neural Network (CNN). Figures 8.3 and 8.4 illustrate the architectures of these two models, respectively.

The client model $\Phi_c$ processes the raw input data and transmits the encoded features to the server, ensuring data protection. This client model is structured with the following layers: A Convolutional Layer (Conv1D) for extracting features from the raw data; an Attention Layer (SE) [29], which refines feature representation by prioritizing the most salient input components; a Batch Normalization Layer (BN) to normalize activations and stabilize the training process; and an Activation Layer

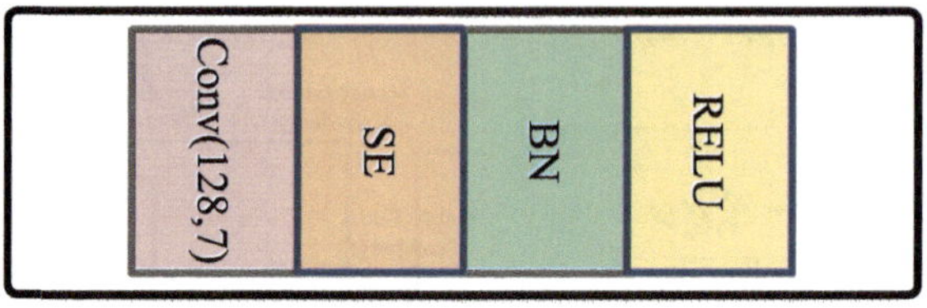


**Fig. 8.3** Neural network architecture of client model

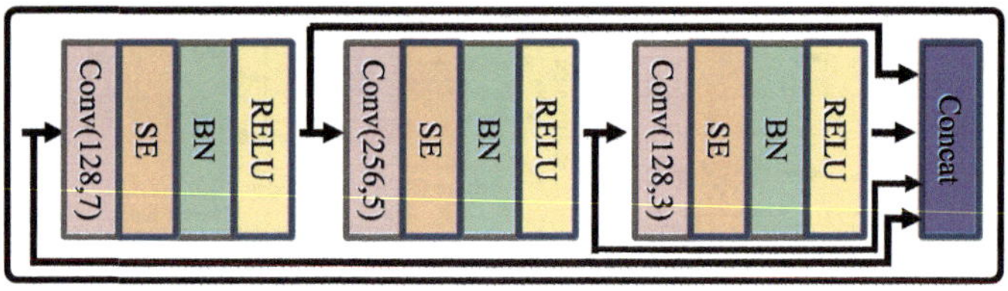


**Fig. 8.4** Neural network architecture of server model

(LeakyReLU) to introduce nonlinearity into the model. It is crucial to note that the encoded features relayed to the server are the outputs from the LeakyReLU activation layer. Given the raw input $X_{raw} \in \Re^{T \times d}$, where $d$ denotes the input dimension and $T$ is the sample window size, the outputs of the SLCNN can be mathematically expressed as

$$\begin{aligned} X_c &= Conv1D(X_{raw}^T), \\ X_{SE} &= SE(X_c), \\ X_{SLCNN} &= LeakyReLU(BN(X_{SE})), \end{aligned} \tag{8.1}$$

where $X_c$, $X_{SE}$, and $X_{SLCNN}$ represent the outputs of the Conv1D, SE, and LeakyReLU layers, respectively. These outputs maintain a consistent shape, $\Re^{F \times L}$, where $F$ is the count of filters in the CNN and $L$ signifies the length of the encoded feature map. To facilitate knowledge transfer, BiFL transmits the extracted features $X_{SLCNN}$, the predicted labels, $Y_p$, and true labels, $Y_t$, to the server.

The inputs to the server model, designated as $\Phi_s$, comprise the local features extracted from the client models. Considering the potentially large number of client models, we utilize a densely connected CNN for the server model to efficiently consolidate information from these client models. The server model is structured with three convolutional blocks, $X_1$, $X_2$, and $X_3$, each mirroring the convolutional block structure of the client model. This architecture enables hierarchical information learning through convolutional operations and dense interconnections. The server model's output is represented as $X_s = [X_f, X_1, X_2, X_3]$, where $[\cdot, \cdot]$ denotes concatenation and $X_f$ refers to the features extracted from the client models.

### 8.2.3 Prototype Method for Class-Imbalanced Data

Class imbalance within training datasets can substantially impede the performance of resultant models, notably affecting the classification accuracy for minority classes. This imbalance issue can be addressed at two distinct levels: the data level and the algorithm level. Data-level strategies typically involve preprocessing the raw data, employing techniques such as resampling or data augmentation, to either amplify the representation of minority classes or diminish the dominance of majority classes [30]. However, these methods frequently necessitate supplemental information, like data distribution, which may conflict with data privacy imperatives, rendering them less suitable for the FL framework [31]. Conversely, algorithm-level strategies prioritize modifying the training algorithm or fine-tuning its parameters [32, 33]. These methods generally require minimal data preprocessing, enhancing their compatibility with FL, although they often demand meticulous tuning of multiple hyperparameters. In this research, we propose a prototype-based method to address the class-imbalance challenge within the FL context. Rather than resampling the raw data, our methodology balances the features via a neural network and adjusts its hyperparameters accordingly. Given the inherent imbalance of the raw data, the latent features extracted by the neural network exhibit similar imbalance. To alleviate this, we initially learn a prototype for each class in the latent space, derived through feature extraction. Subsequently, we construct a classifier based on these learned prototypes. Critically, this method operates without requiring knowledge of the raw data distribution, thereby preserving privacy. To further optimize the efficacy of our approach, we integrate an attention-oriented prototype learning method [34], which accentuates pertinent information while disregarding irrelevant features.

A prototype is essentially a vector that encapsulates each class. The most straightforward method to compute a prototype involves averaging all features within the latent space:

$$C^i = \frac{1}{n^i} \sum_{k=1}^{n^i} X^{i,k}, \tag{8.2}$$

where $C^i \in \mathbb{R}^{1\times H}$ is the prototype for class $i \in [1, \ldots, n_c]$, $n_c$ is the total number of classes, $n^i$ is the number of instances in the deep latent space for class $i$, and $X$ represents the features of the samples in the deep latent space.

While prototype calculation based on Eq. (8.2) is effective, emphasizing relevant features while disregarding irrelevant ones is crucial. To address this, we employ an attentional prototype (AP) [34], defined as follows. Let $X_k = [x_1, \ldots, x_m] \in \Re^{m\times dl}$ represent the matrix of latent features for class $k$, where $m$ is the total number of data samples with class label $k$, and $dl$ is the output dimension of both client and server models.

$$e_k = sigmoid(U_k^T tanh(V_k X_k^T)), \tag{8.3}$$

where $U_k \in \Re^{w \times 1}$ and $V_k \in \Re^{w \times dl}$ represent the trainable parameters of the prototype attention module, with $w$ denoting the dimensionality of the two attention parameters.

The attention weight can then be computed as

$$\alpha_k = \frac{exp(e_k)}{\sum_{j=1}^{m} exp(e_k^j)}. \tag{8.4}$$

Once the attention weight is computed, the attentional prototype for class $k$ can be obtained as follows:

$$C_k = \alpha_k * X_k. \tag{8.5}$$

In this research, we introduce distinct parameters, $U_k$ and $V_k$, for each class to account for the varying attention levels that different classes may exhibit within their feature spaces. The imbalance in feature distribution is mitigated through the application of Eq. (8.5).

**Algorithm 4** The training process of the server model

**Input:** Training epoch $E_s$, communication rounds $R$, the number of clients $K$, server model $\Phi_s$

```
Server:
  Initialize the model parameters W_s of the server model Φ_s
  for each communication round r to R do
    for each client k to K do
      X_c^k, y_t^k ← Client(r, k)
      F[k] ← X_c^k
      Y_t[k] ← y_t^k
    end
    for e = 0; e < E_s; e = e + 1 do
      for each client k to K do
        X_s^k ← Φ_s(F[k])                         /* features of the k-th client */
        for each class i do
          X_i^k ← X_s^k(idx = i)                   /* matrix for class i */
          e_i^k ← sigmoid((U_i^k)^T tanh(V_i^k (X_i^k)^T))
          α_i^k ← exp(e_i^k) / Σ_{j=1}^{m} exp((e_i^k)^j)
          C_i^k ← α_i^k * X_i^k                    /* prototypes */
        end
        p_s^k ← exp(−dist(Φ_s(X), C_i^k)) / Σ_n exp(−dist(Φ_s(X), C_n))   /* probability */
        Sending p_s^k to the k-th client
        l_s ← L_CE(Y_t[k], p_s^k)                  /* CE loss (Eq. (8.6)) */
        W_s ← W_s − η_s ∇l_s(W_s)                  /* parameter update */
      end
    end
  end
```

### 8.2.4 Knowledge Transfer Enhanced by KD

To train the classifier, we employ the binary cross-entropy loss defined as

$$L_{CE} = -\frac{1}{N}\sum_{i=1}^{N} y_i \log(p(y_i)) + (1 - y_i)\log(1 - p(y_i)), \tag{8.6}$$

where $y_i$ is the true value for the $i$-th class, $p(\cdot)$ is the estimated probability for the $i$-th class, and $N$ is the total number of samples. To estimate the probability for the $i$-th class, we utilize the squared Euclidean distance between the $i$-th class prototype and a given time series, as defined by

$$p(y_i) = \frac{exp(-dist(\Phi(X), C_i))}{\sum_k exp(-dist(\Phi(X), C_k))}, \tag{8.7}$$

where $\Phi$ is the neural network (network used in both client and server models) and $C_i$ is the prototype of the $i$-th class.

In this study, each client transmits its learned knowledge to the server, which subsequently relays the aggregated information back to the clients. Knowledge Distillation (KD) facilitates knowledge transfer from the server model to each client model. Given the potentially large number of clients, knowledge transfer from clients to the server does not utilize KD, in contrast to approaches such as [35], which rely on a public dataset. KD is directly implemented by comparing the predicted outputs of client models with those of the server model.

$$L_{KD} = KL(p_c||p_s) = \frac{1}{K}\sum_{i}^{N} p_c log(\frac{p_c}{p_s}), \tag{8.8}$$

where $p_c$ and $p_s$ are the predicted values of the client model and server model, respectively.

BiFL employs the subsequent loss function, which integrates the two aforementioned losses for training:

$$L_{all} = \alpha L_{CE} + (1 - \alpha)L_{KD}, \tag{8.9}$$

where $\alpha$ is the hyperparameter balancing the cross-entropy loss and KD loss, set to 0.9 in our experiments.

Algorithms 4 and 5 delineate the training procedures for client and server models, respectively.

**Algorithm 5** The training process of the client model

**Input:** Training epoch $E_c$, $\alpha$, client model $\Phi_c$

```
Client(r, k):
  Initialize model parameters W_c of model Φ_c
  for e = 0; e < E_c; e = e + 1 do
                                        /* features from raw sensor data */
    for each class i do
      X_i ← X_c(idx = i)                          /* matrix for class i */
      e_i ← sigmoid(U_i^T tanh(V_i X_i^T))
      α_i ← exp(e_i) / Σ_{j=1}^{m} exp(e_i^j)
      C_i ← α_i * X_i                        /* prototype for each class */
    end
    p_c ← exp(−dist(Φ_c(X), C_i)) / Σ_k exp(−dist(Φ_c(X), C_k))   /* probability */
    l_ce ← L_CE(y_t, p_c)                /* classification loss (Eq. (8.6)) */
    l_kd ← L_KD(p_c, p_s^k)                      /* KD loss (Eq.(8.8)) */
    l_c ← α * l_ce + (1 − α)l_kd               /* total loss (Eq. (8.9)) */
    W_c ← W_c − η_c ∇l_c(W_c)                       /* parameter update */
  end
  return X_c, y_t
```

### 8.2.5 *Knowledge Protection Mechanism*

As previously discussed, the exchange of learned feature maps and labels between clients and the central server over the network is essential; however, this process is susceptible to cyber-attacks, potentially leading to data leakage. To mitigate this concern, we propose a knowledge protection mechanism, illustrated in Fig. 8.2. This mechanism obfuscates feature maps and labels, reducing their correlation with original data, thus providing protection against adversarial reconstruction [36].

Consider $K$ client models (data owners) $l_1, l_2, \ldots, l_K$ and a global server model $S$. Each client uses its dataset $X^i_{raw}$ $(i \in 1, \ldots, K)$ to train a local deep neural network $\Omega_i$ $(i \in 1, \ldots, K)$. The dataset $X^i_{raw} \in \Re^{n \times T_i \times d_i}$ comprises $n$ training samples, where each sample's shape is client specific. To expedite training, these $n$ samples are reorganized into $B_i$ batches. Note that the values of $T_i$ and $d_i$ $(i \in 1, \ldots, K)$ are confidential and not shared among clients. Let the raw learned feature map of the $i$-th client be $X^i_h \in \Re^{B_i \times F \times L}$, where $F$ and $L$ are consistent across all clients. The knowledge protection mechanism is then mathematically formulated:

$$X^i_{Enc} = \underbrace{Enc(\cdots Enc(Enc}_{e-1}(X^i_h, X^i_{h1}, \beta_1), X^i_{h2}, \beta_2), X^i_{h3}, \beta_3), \\ \cdots, X^i_{he}, \beta_e) \tag{8.10}$$

$$Y^i_{Enc} = \underbrace{Enc(\cdots Enc(Enc}_{e-1}(Y^i, Y^i_1, \beta_1), Y^i_2, \beta_2), Y^i_3, \beta_3), \cdots, Y^i_e, \beta_e) \tag{8.11}$$

$$Enc(X_1, X_2, \beta) = \beta \cdot X_1 + (1 - \beta) \cdot X_2, \tag{8.12}$$

where $X^i_{hj}$ and $Y^i_j$ ($j \in \{1, \ldots, e\}$) represent the encrypted feature map and label, respectively, and $\beta_j \sim Beta(1, 1)$ is the encoding factor. The *Enc* operation is defined as a linear combination of two inputs, weighed by $\beta$.

With the proposed data protection mechanism, each client securely transmits the encoded feature map to the server. Ensuring consistent dimensionality of feature maps sent to the server is crucial, potentially corresponding to the output of any hidden layer. Given client model heterogeneity in BiFL, where architectures may vary, we assume participants lack access to each other's model structures. This assumption further reduces the risk of feature-based inference attacks. In this research, knowledge exchange between clients and server occurs solely during training. Consequently, data reconstruction risk is minimized, as attacker access is limited to evolving, untrained feature maps, not fully trained maps directly representing raw data [28]. Feature maps are discarded after the server generates the global model.

BiFL's design mitigates peer inference attack risks between clients via a non-sharing architecture, maintaining data confidentiality for each client. However, in scenarios where a client might attempt inference or pattern memory attacks on others, additional protection mechanisms—such as secure multi-party computation, homomorphic encryption, blockchain, or differential privacy—can be implemented to further enhance data security.

### *8.2.6 Computational Complexity Analysis*

Computationally intensive components of client and server models include the Conv1D layer, SE layer, and prototype learning module. Computational complexity is $O(F \cdot k \cdot d \cdot T)$ for Conv1D, $O(F^2)$ for SE, and $O(w \cdot dl \cdot m)$ for prototype learning. Here, $F$ is the number of filters, $k$ is filter size, $d$ is the number of features, and $T$ is time-series length. Parameters $w$, $m$, and $dl$ are associated with the attentional prototype module, where $w$ is the size of trainable parameters, $m$ is the number of feature samples, and $dl$ is feature dimension. Knowledge protection mechanism complexity is $O(e)$, negligible as $e$ is small ($< 4$). Server and client model combined complexity is: $O(n_s \cdot (F_s \cdot k_s \cdot d_s \cdot T_s + F_s^2) + w_s \cdot dl_s \cdot m_s)$ and $O(F_c \cdot k_c \cdot d_c \cdot T_c + F_c^2 + w_c \cdot dl_c \cdot m_c)$, where $n_s$ is server model layers, and subscripts $s$ and $c$ denote server and client models. Given data size $D$ for each client, BiFL framework overall complexity is: $O[R \cdot (E_s \cdot K \cdot D \cdot ((n_s \cdot (F_s \cdot k_s \cdot d_s \cdot T_s + F_s^2) + w_s \cdot dl_s \cdot m_s) + E_c \cdot K \cdot D \cdot (F_c \cdot k_c \cdot d_c \cdot T_c + F_c^2 + w_c \cdot dl_c \cdot m_c)))]$, where

$R$ is communication rounds, $K$ is clients' number, and $E_s$ and $E_c$ are training epochs for server and clients.

## 8.3 Experiment

Models were implemented using PyTorch (version 1.8.0). All experiments were executed on a server with a 16GB Tesla T4 GPU. Training hyperparameters were: Adam optimizer with a $1e-3$ learning rate, training epochs $E_c = 5$ and $E_s = 10$ for client and server, and total communication rounds $R = 20$.

### 8.3.1 *Experimental Settings and Data*

The experimental dataset originated from two wind farms in Shanxi and Henan Provinces, China, approximately 700 kilometers apart in northern China. Twenty wind turbines were simulated as clients, 10 per wind farm, with a central server for data aggregation. Data collection for the first wind farm spanned from February 11, 2019 to February 26, 2019; the second wind farm's data were collected from February 12, 2019 to February 26, 2019. Temporal resolution was set to 30 seconds.

Wind turbine operation experts identified 16 variables associated with blade icing, detailed in Table 8.1, and classified data into "icing" and "normal". Raw data exhibited quality issues like noise, outliers, and missing values, addressed through data cleaning before modeling. To mitigate varying scales, data were normalized to scalar values and raw time-series data were segmented into fixed-length windows.

60% of data was allocated for training per client and 40% for testing. To evaluate global model robustness, three class-imbalance ratios were tested, specifically the ratio of normal to icing class samples. Imbalance ratio $\rho$ is $\rho = \frac{N_{\text{normal}}}{N_{\text{Icing}}}$, where $N_{\text{normal}}$ and $N_{\text{Icing}}$ are normal and icing sample counts. Test data imbalance ratio was set to 10:1.

### 8.3.2 *Evaluation Metrics*

Two metrics are used: $F_\beta$ for classification performance and $BA$ (balanced accuracy) for imbalanced learning, defined as

$$F_\beta = \frac{(1+\beta^2) \times Precision \times Recall}{\beta^2 Precision + Recall}, \tag{8.13}$$

$$Precision = \frac{TP}{TP+FP}, \quad Recall = \frac{TP}{TP+FN}, \tag{8.14}$$

**Table 8.1** SCADA data specification

| No. | Variable name | Description |
|---|---|---|
| 1 | wind_speed | Wind speed |
| 2 | wind_direction | Wind direction |
| 3 | generator_speed | Generator speed |
| 4 | power | Active power |
| 5 | yaw_position | Yaw position |
| 6 | pitch1_angle | Angle of pitch 1 |
| 7 | pitch2_angle | Angle of pitch 2 |
| 8 | pitch3_angle | Angle of pitch 3 |
| 9 | pitch1_speed | Speed of pitch 1 |
| 10 | pitch2_speed | Speed of pitch 2 |
| 11 | pitch3_speed | Speed of pitch 3 |
| 12 | environment_temp | Environment temperature |
| 13 | internal_temp | Internal temperature of nacelle |
| 14 | pitch1_moto_tmp | Temperature of pitch motor 1 |
| 15 | pitch2_moto_tmp | Temperature of pitch motor 2 |
| 16 | pitch3_moto_tmp | Temperature of pitch motor 3 |

$$\mathrm{BA} = \frac{1}{2}\left(\frac{TP}{TP + FN} + \frac{TN}{TN + FP}\right), \tag{8.15}$$

where $TP$, $FP$, $FN$, and $TN$ are true positive, false positive, false negative, and true negative, respectively. $\beta$ in $F_\beta$ was set to 2.

For $K$ client models, average metric values are

$$mF_\beta = \frac{1}{K}\sum_{i}^{K} F_\beta^i, \quad mBA = \frac{1}{K}\sum_{i}^{K} BA^i. \tag{8.16}$$

To minimize randomness effects, values from all server training rounds per communication cycle were averaged.

### 8.3.3 Comparison with State-of-the-Art FL Methods

Two FL frameworks are evaluated: (1) FedAvg (Federated Averaging) [37], a classic FL framework. FedAvg utilizes multiple clients and a central server. Clients train local models with their data and upload trained models to the server. The server initializes the network and aggregates client models using weighted averaging. FedAvg requires identical network architectures for server and clients. (2) FLGKT (Federated

**Table 8.2** Performance comparison with federated learning methods

| Method | $\rho = 20:1$ | | $\rho = 50:1$ | | $\rho = 100:1$ | |
|---|---|---|---|---|---|---|
| | $mF_\beta$ | $mBA$ | $mF_\beta$ | $mBA$ | $mF_\beta$ | $mBA$ |
| FedAvg | 45.56 | 70.01 | 38.64 | 64.37 | 37.95 | 63.74 |
| FLGKT | **80.14** | **91.05** | 61.28 | 80.70 | 58.02 | 76.43 |
| *Ours* | 77.72 | 90.25 | **65.76** | **84.77** | **64.76** | **83.82** |

Learning as Group Knowledge Transfer) [28], a state-of-the-art FL framework for knowledge transfer from clients to server. Clients use lighter models than the server to reduce computation.

Sampling window size is 128 (approximately 11 minutes). In BiFL, CNN filter number is 128. Parameter $\alpha$ is 0.9 and parameter $m$ is 2. Table 8.2 shows results, with best values in bold. BiFL performs better in most scenarios, except $\rho = 20:1$, where it is second to FLGKT. This is due to FLGKT's more complex client and server models. At $\rho = 50:1$, our model improves $mF_\beta$ and $mBA$ by 41.24% and 24.07% over FedAvg, comparable to FLGKT. At $\rho = 100:1$, our model further improves, with $mF_\beta$ and $mBA$ increasing by 41.40% and 23.96% over FedAvg. Compared to FLGKT, our model improves $mF_\beta$ and $mBA$ by 10.41% and 8.82%. As imbalance ratio increases, accuracy of all three methods decreases.

### *8.3.4 Comparison with State-of-the-Art Class-Imbalance Methods*

BiFL is compared with six class-imbalanced algorithms: **Focal** (focal loss): A common loss function for class imbalance. We set $\gamma = 1$ and use class weights from [32]. **WCE** (Weighted Cross-Entropy): Standard cross-entropy loss extension with class-specific weights for imbalances. **CB** (Class Balance): Uses the effective sample number, considering data overlap. Class-balance loss is based on effective sample numbers per class [33]. **GHMC** (Gradient Harmonizing Mechanism Classification): Addresses disharmony in imbalanced classification [38]. **LDAM** (Label-Distribution-Aware Margin): Replaces standard cross-entropy loss, combinable with re-weighting/resampling [39]. **Balance**: Data-level algorithm undersampling the majority class for imbalance [30].

Table 8.3 shows our model outperforms others in $mF_\beta$ and $mBA$, surpassing Focal, WCE, CB, GHMC, and LDAM. WCE is second in all scenarios, LDAM consistently lowest in $mF_\beta$. Our model improves $mF_\beta$ and $mBA$ by 41.96% and 23.00% over WCE at $\rho = 20:1$. At $\rho = 50:1$ and $\rho = 100:1$, our model increases $mF_\beta$ by 32.34% and 32.52%, and $mBA$ by 18.56% and 18.73% over WCE. Compared to CB, our model is competitive in $mF_\beta$ but slightly lower in $mBA$ at $\rho = 50:1$. $mBA$ emphasizes overall accuracy, $mF_\beta$ specifically assesses icing class prediction

**Table 8.3** Performance comparison with state-of-the-art class-imbalanced methods

| Method | $\rho = 20:1$ | | $\rho = 50:1$ | | $\rho = 100:1$ | |
|---|---|---|---|---|---|---|
| | $mF_\beta$ | $mBA$ | $mF_\beta$ | $mBA$ | $mF_\beta$ | $mBA$ |
| Focal | 25.74 | 55.38 | 21.68 | 53.18 | 24.82 | 54.41 |
| WCE | 45.11 | 69.49 | 44.49 | 69.04 | 43.70 | 68.12 |
| CB | 44.55 | 68.88 | 43.91 | 68.37 | 43.09 | 67.42 |
| GHMC | 27.94 | 51.21 | 26.92 | 50.12 | 27.11 | 50.27 |
| LDAM | 18.96 | 52.35 | 16.73 | 51.38 | 16.31 | 50.86 |
| Balance | 72.15 | 88.83 | 65.66 | **85.04** | 62.22 | 82.36 |
| *Ours* | **77.72** | **90.25** | **65.76** | 84.77 | **64.76** | **83.82** |

accuracy. Our approach is superior to Balance, especially for blade icing detection. Data balancing across clients is also time and resource intensive.

### 8.3.5 Impact of Knowledge Protection Mechanism

This section examines the performance impact of the knowledge protection mechanism. We vary parameter $m$ from 2 to 4 and compare results with **No_KPM**, where the mechanism is omitted from BiFL.

Table 8.4 shows knowledge protection mechanism inclusion degrades performance for $\rho = 20:1$ and $\rho = 50:1$. At $\rho = 100:1$, $m$=*3* and $m$=*4* show slight performance improvements. Further increasing $m$ makes performance changes irregular. Contrary to expectations, no consistent downward trend in performance occurs as imbalance ratio $\rho$ increases with knowledge protection. This is because the encoding process distorts raw features, negatively impacting prototypes and causing deviations, affecting final results.

**Table 8.4** Performance impacts of knowledge protection mechanism

| Method | $\rho = 20:1$ | | $\rho = 50:1$ | | $\rho = 100:1$ | |
|---|---|---|---|---|---|---|
| | $mF_\beta$ | $mBA$ | $mF_\beta$ | $mBA$ | $mF_\beta$ | $mBA$ |
| $m = 2$ | 77.72 | 90.25 | 65.76 | 84.77 | 64.76 | 83.82 |
| $m = 3$ | 68.07 | 86.13 | 72.19 | 88.27 | **74.43** | **88.44** |
| $m = 4$ | 65.21 | 84.85 | 71.78 | **88.42** | 72.01 | 87.86 |
| *No_KPM* | **86.28** | **92.61** | **75.89** | 86.35 | 68.71 | 82.24 |

**Table 8.5** Ablation study

| Method | $\rho = 20:1$ | | $\rho = 50:1$ | | $\rho = 100:1$ | |
|---|---|---|---|---|---|---|
| | $mF_\beta$ | $mBA$ | $mF_\beta$ | $mBA$ | $mF_\beta$ | $mBA$ |
| No_AP | 34.79 | 60.35 | 32.29 | 57.34 | 35.22 | 60.14 |
| No_AttP | 52.80 | 75.94 | 51.17 | 74.68 | 44.63 | 69.52 |
| No_SE | 60.03 | 81.29 | 59.07 | 80.79 | 58.70 | 79.97 |
| *Ours* | **77.72** | **90.25** | **65.76** | **84.77** | **64.76** | **83.82** |

### *8.3.6 Ablation and Sensitivity Analysis*

We perform an ablation study to assess AP, SE, and KD module contributions. Four BiFL variants are evaluated: (1) **No_AP**, omitting the AP module; (2) **No_AttP**, removing the attention module from AP (prototype computed using Eq. (8.2)); 3) **No_SE**, excluding the SE module.

Table 8.5 shows results. Performance degradation is most pronounced without the AP module in all cases. $mF_\beta$ decreases by 42.93%, 33.47%, and 29.54%, and $mBA$ drops by 29.90%, 27.43%, and 23.68%. Excluding the SE module decreases $mF_\beta$ by 17.69%, 6.69%, and 6.06%, and $mBA$ by 8.96%, 3.98%, and 3.58%. Comparing No_AP and No_AttP variants shows improvements in both metrics across all cases, highlighting attention module effectiveness in prototype learning. Proposed modules significantly contribute to model performance, individually and combined.

To assess input window size and $\alpha$ influence, we conducted sensitivity analysis, varying window size from 32 to 256 and $\alpha$ from 0.15 to 0.55. Results are in Figs. 8.5 and 8.6. Figure 8.5 shows both metrics ($mF_\beta$ and $mBA$) increase almost linearly with window size, optimal at 256. Figure 8.6 shows performance slightly declines as $\alpha$ increases for $\rho = 20:1$ and $\rho = 50:1$. For $\rho = 100:1$, performance fluctuates marginally, indicating subtle $\alpha$ effect. Equation (8.9) shows smaller $\alpha$ reduces cross-entropy loss influence, amplifying KD effect. This supports knowledge transfer's positive impact from server to client models.

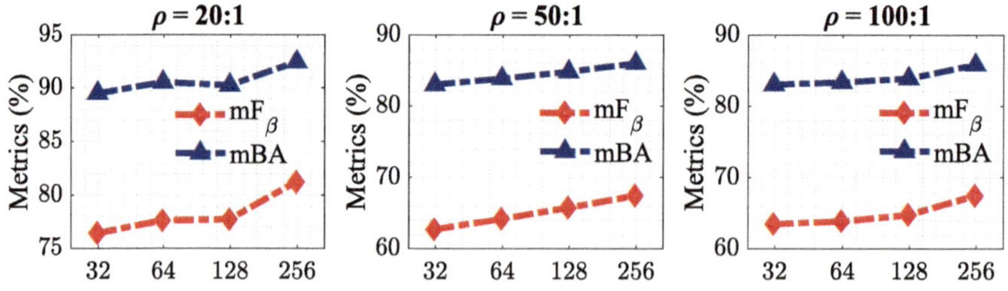


**Fig. 8.5** Sensitivity analysis of window size

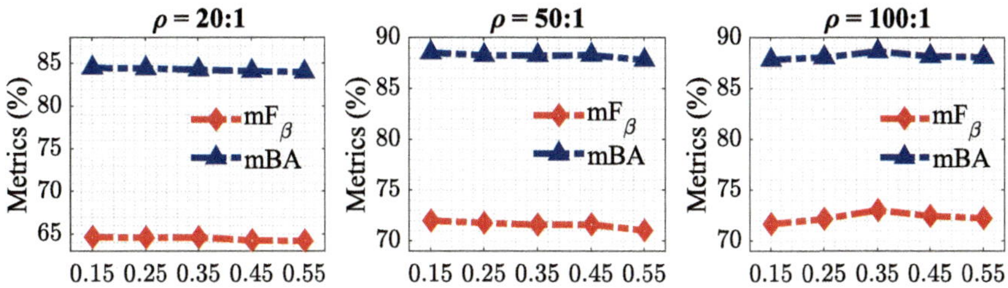


**Fig. 8.6** Sensitivity analysis of $\alpha$

## 8.4 Conclusion

This research introduces BiFL, a class-imbalanced, heterogeneous FL model for wind turbine blade icing detection. Given blade icing's strong dependence on local conditions and turbine characteristics, BiFL uses transfer learning to enhance robustness and generalization. A prototype-based method addresses class imbalance in training data. Comprehensive experiments compared BiFL to traditional FL models and state-of-the-art class-imbalance methods, demonstrating superior performance. Ablation studies confirmed BiFL component effectiveness, and sensitivity analysis evaluated key hyperparameter impact.

Future research directions include further mitigating data breach risks, as clients upload extracted features to the server. Differential privacy and blockchain could secure uploaded data and protect against network attacks. More effective icing severity assessment is needed, potentially interpreting output probabilities or using domain knowledge. Depthwise convolutions and efficient operations could design lightweight CNNs, reducing complexity while enabling deeper networks for improved feature learning.

## References


1. Wei K, Yang Y, Zuo H, Zhong D (2020) A review on ice detection technology and ice elimination technology for wind turbine. Wind Energy 23(3):433–457
2. Laakso T, Holttinen H, Ronsten G, Tallhaug L, Horbaty R, Baring-Gould I, Lacroix A, Peltola E, Tammelin B (2003) State-of-the-art of wind energy in cold climates. IEA Annex XIX 24:53
3. Zeng J, Song B (2017) Research on experiment and numerical simulation of ultrasonic de-icing for wind turbine blades. Renew Energy 113:706–712
4. Corradini ML, Cristofaro A, Pettinari S (2016) "A model-based robust icing detection and estimation scheme for wind turbines. In: European control conference (ECC). IEEE, pp 1451–1456
5. Jolin N, Bolduc D, Swytink-Binnema N, Rosso G, Godreau C (2019) Wind turbine blade ice accretion: a correlation with nacelle ice accretion. Cold Regions Sci Technol 157:235–241
6. Shu L, Qiu G, Hu Q, Jiang X, McClure G, Yang H (2018) Numerical and field experimental investigation of wind turbine dynamic de-icing process. J Wind Eng Indus Aerodyn 175:90–99

7. Yin C, Zhang Z, Wang Z, Guo H (2016) Numerical simulation and experimental validation of ultrasonic de-icing system for wind turbine blade. Appl Acoust 114:19–26
8. Jiménez AA, Márquez F, Moraleda VB, Muñoz C (2019) Linear and nonlinear features and machine learning for wind turbine blade ice detection and diagnosis. Renew Energy 132:1034–1048
9. Liu Y, Cheng H, Kong X, Wang Q, Cui H (2019) Intelligent wind turbine blade icing detection using supervisory control and data acquisition data and ensemble deep learning. Energy Sci Eng 7(6):2633–2645
10. Yuan B, Wang C, Jiang F, Long M, Yu PS, Liu Y (2019) Waveletfcnn: a deep time series classification model for wind turbine blade icing detection. arXiv preprint arXiv:1902.05625
11. Cheng X, Shi F, Zhao M, Li G, Zhang H, Chen S (2021) Temporal attention convolutional neural network for estimation of icing probability on wind turbine blades. IEEE Trans Indus Electron 69(6):6371–6380
12. Goncalves C, Bessa RJ, Pinson P (2021) Privacy-preserving distributed learning for renewable energy forecasting. IEEE Trans Sustain Energy 12(3):1777–1787
13. Wang L, Xu S, Wang X, Zhu Q (2021) Addressing class imbalance in federated learning. In: Proceedings of the AAAI conference on artificial intelligence, vol 35, no 11, pp 10 165–10 173
14. Zhang X, Fang F, Wang J (2020) Probabilistic solar irradiation forecasting based on variational bayesian inference with secure federated learning. IEEE Trans Indus Inform
15. Venkataramanan V, Kaza S, Annaswamy AM (2021) Der forecast using privacy preserving federated learning. arXiv preprint arXiv:2107.03248
16. Wang Y, Bennani IL, Liu X, Sun M, Zhou Y (2021) Electricity consumer characteristics identification: a federated learning approach. IEEE Trans Smart Grid
17. Kariniotakis G, Camal S, van der Meer D, Pinson P, Giebel G, Han L, Bessa R, Libois Q, Cassas M, Alonzo B et al (2021) Research directions and results in the smart4res project for improving renewable energy forecasting. In: Wind energy science conference (WESC)
18. Phoenix Project (2020), https://phoenix-h2020.eu/
19. Li T, Sahu AK, Zaheer M, Sanjabi M, Talwalkar A, Smith V (2020) Federated optimization in heterogeneous networks. MLSys
20. Arivazhagan MG, Aggarwal V, Singh AK, Choudhary S (2019) Federated learning with personalization layers. arXiv preprint arXiv:1912.00818
21. Liang PP, Liu T, Ziyin L, Salakhutdinov R, Morency L-P (2020) Think locally, act globally: Federated learning with local and global representations. Adv Neural Inform Process Syst
22. Ghosh A, Chung J, Yin D, Ramchandran K (2020) An efficient framework for clustered federated learning. In: Advances in neural information processing systems
23. Sattler F, Müller K-R, Samek W (2020) Clustered federated learning: model-agnostic distributed multitask optimization under privacy constraints. IEEE Trans Neural Networks Learn Syst
24. Li D, Wang J (2020) Fedmd: Heterogenous federated learning via model distillation.In: Advances in neural information processing systems
25. Long G, Shen T, Tan Y, Gerrard L, Clarke A, Jiang J (2021) Federated learning for privacy-preserving open innovation future on digital health. arXiv preprint arXiv:2108.10761
26. He C, Annavaram M, Avestimehr S (2020) Fednas: federated deep learning via neural architecture search. In: CVPR 2020 workshop on neural architecture search and beyond for representation learning
27. Singh I, Zhou H, Yang K, Ding M, Lin B, Xie P (2020) Differentially-private federated neural architecture search. In: FL-international conference on machine learning workshop
28. He C, Annavaram M, Avestimehr S (2020) Group knowledge transfer: federated learning of large cnns at the edge. In: Advances in neural information processing systems, vol 33. Curran Associates, Inc., pp 14 068–14 080
29. Hu J, Shen L, Sun G (2018) Squeeze-and-excitation networks. In: Proceedings of the IEEE conference on computer vision and pattern recognition, pp 7132–7141
30. Leevy JL, Khoshgoftaar TM, Bauder RA, Seliya N (2018) A survey on addressing high-class imbalance in big data. J Big Data 5(1):1–30

31. Wang L, Xu S, Wang X, Zhu Q (2020) Towards class imbalance in federated learning. arXiv preprint arXiv:2008.06217
32. Lin T-Y, Goyal P, Girshick R, He K, Dollár P (2017) Focal loss for dense object detection. In: Proceedings of the IEEE international conference on computer vision, pp 2980–2988
33. Cui Y, Jia M, Lin T-Y, Song Y, Belongie S (2019) Class-balanced loss based on effective number of samples. In: Proceedings of the IEEE conference on computer vision and pattern recognition, pp 9268–9277
34. Zhang X, Gao Y, Lin J, Lu C-T (2020) Tapnet: multivariate time series classification with attentional prototypical network. In: AAAI, pp 6845–6852
35. Zhuang W, Wen Y, Zhang X, Gan X, Yin D, Zhou D, Zhang S, Yi S (2020) Performance optimization of federated person re-identification via benchmark analysis. In: Proceedings of the 28th ACM international conference on multimedia, pp 955–963
36. Xiao D, Yang C, Wu W (2021) Mixing activations and labels in distributed training for split learning. IEEE Trans Parallel Distrib Syst
37. McMahan B, Moore E, Ramage D, Hampson S, y Arcas BA (2017) Communication-efficient learning of deep networks from decentralized data. In: Artificial intelligence and statistics. PMLR, pp 1273–1282
38. Li B, Liu Y, Wang X (2019) Gradient harmonized single-stage detector. Proc AAAI Conf Artif Intell 33(01):8577–8584
39. Cao K, Wei C, Gaidon A, Arechiga N, Ma T (2019) Learning imbalanced datasets with label-distribution-aware margin loss. In: Advances in neural information processing systems, pp 1567–1578

# Chapter 9
# Data-Efficient Heterogeneous Federated Convolutional Neural Network for Blade Icing Detection

## 9.1 Introduction

Wind energy, a readily accessible and cost-free resource characterized by exceptionally low pollutant emissions throughout its lifecycle, presents a compelling solution to the growing global need for clean energy sources [1]. The deployment of wind turbines has significantly increased in recent decades, particularly in areas near urban centers. Alongside hydropower, wind energy consumption has become a leading contributor in the renewable energy sector [2]. However, wind farms are predominantly situated in high-altitude regions, where reduced ambient temperatures expose wind turbine blades to harsh cold climates, especially during winter. This exposure frequently leads to ice accumulation on the blades. Ice accretion can severely degrade power generation, potentially causing annual reductions as high as 30%, and can also pose considerable safety hazards to adjacent infrastructure and equipment [3]. Consequently, the imperative for effective icing prediction has become increasingly prominent in relevant research domains.

Various methodologies, including direct, indirect, model-driven, and data-driven approaches, have been explored for classifying turbine blade icing conditions. Direct methods analyze the physical characteristics of ice, utilizing techniques such as ultrasonic damping [4], piezoelectric sensors [5, 6], and temperature variation detection [7]. Nevertheless, many direct methods encounter limitations in practical deployment, and some struggle to provide timely detection of icing conditions [8, 9]. Indirect methods, which often necessitate supplementary external instruments, include techniques like dual anemometers [10], visibility and cloud base height measurement [11], and monitoring of engine room vibrations and power curves [12–14]. However, these indirect approaches are susceptible to errors from machine malfunctions and frequently require integration with weather forecasts to attain acceptable accuracy. Model-driven approaches employ statistical or mathematical models, often in conjunction with weather predictions and computational fluid dynamics (CFD), to forecast icing-induced energy losses in wind turbines operating in cold climates [15, 16].



X. Cheng et al., *Computational Methods for Blade Icing Detection of Wind Turbines*, Engineering Applications of Computational Methods 24,
https://doi.org/10.1007/978-981-96-6763-5_9

However, model-driven techniques rely on specific assumptions about icing conditions and often necessitate external tools, such as wind tunnels, for experimental validation. Furthermore, a thorough understanding of the icing process is crucial for precise modeling [17].

In recent years, data-driven methodologies, encompassing machine learning and deep learning techniques, have garnered increasing interest for icing prediction. Chen et al. presented a deep neural network incorporating a bypass component to learn deep representations from imbalanced data for wind turbine fault detection [18]. Yuan et al. introduced a wavelet-enhanced autoencoder for identifying icing conditions [19], while Tian et al. developed a multilevel convolutional recurrent model, integrating wavelet transformation for turbine icing detection [20]. Tong et al. proposed an adaptive weighted kernel extreme learning machine algorithm that incorporates sample distribution properties [21]. Moreover, Cheng et al. designed a temporal attention convolutional model for icing estimation [22], and Wang et al. presented a wavelet-based multi-scale long short-term memory network for wind turbine blade icing detection [23]. Xiao et al. introduced a selective deep ensemble model utilizing the group method of data handling technique [24]. Most existing data-driven studies train models using a centralized paradigm, aggregating turbine data from diverse wind farms at a central location, such as a cloud server.

While effective, traditional centralized approaches, which consolidate data in a unified cloud environment for model training, impose significant demands on storage capacity and computational resources. Additionally, sufficient network bandwidth between the central server and each wind turbine is essential to facilitate data transfer. Furthermore, data centralization raises potential privacy concerns, even though machine data may not possess the same sensitivity as personal information. Nevertheless, wind farm operators prioritize data security and restrict external data access, leading to the formation of isolated "data islands" across wind farms. Federated Learning (FL) has emerged as a promising solution to these challenges [25]. The concept of federated learning (FL) was initially proposed by McMahan et al. [25] as a distributed model training paradigm. In FL, participants train local models independently using their private data, and, subsequently, a global model is aggregated on a central server without necessitating the upload of raw data. By shifting the training process away from a central server to distributed client devices, FL enables each client to maintain data locality. Only model weights are periodically updated through synchronization with clients, significantly reducing computational burdens and enhancing data privacy compared to centralized approaches. Hard et al. demonstrated the applicability of FL by training a recurrent neural network language model to predict the subsequent word on smartphone virtual keyboards [26]. In the medical domain, Sheller et al. applied FL, demonstrating its capacity to enable robust learning across datasets without compromising sensitive patient data, and to foster large-scale inter-institutional collaboration while adhering to data protection regulations, such as the European Union's General Data Protection Regulation (GDPR) [27], aimed at safeguarding data privacy [28]. Wang et al. employed FL to develop a model for identifying electricity consumer profiles [29], while Cheng et al. introduced a blockchain-based federated model for turbine blade icing detection [30]. Pei et al.

proposed a personalized FL framework for anomaly detection in network traffic [31], and Cheng et al. further advanced FL by implementing a heterogeneous client–server model to address turbine blade icing detection [32]. Additionally, Arikumar et al. developed an FL-based approach for identifying individual movement patterns using wearable devices [33].

Applying federated learning (FL) for wind turbine blade icing prediction introduces several notable challenges: (1) Feature relevance is paramount for achieving high model quality, particularly in FL. Wind turbines are equipped with numerous sensors, yet not all sensors contribute effectively to blade icing detection. For example, temperature data may be highly relevant, whereas other sensor data might provide minimal information. Training models without feature selection could impose an unnecessary computational load. (2) Wind turbines typically operate continuously, with shutdowns due to ice accumulation being infrequent. Consequently, data representing normal operation vastly outweighs data indicating icing events, leading to significant class imbalance. Training on such imbalanced data without mitigation strategies can bias the model towards the dominant class, hindering its ability to detect rare icing events. (3) Although FL decentralizes training by distributing it across client devices, local training on each client remains essentially centralized. Local clients exhibit varying convergence rates, and their communication status differs considerably, resulting in inconsistent participation in communication rounds. As a result, when the central server aggregates parameters from participating client models, clients with slower convergence or irregular participation can disrupt the convergence of the global model, slowing down the overall training process. Despite these advancements in FL and its inherent potential, the application of FL to wind turbine icing prediction remains relatively underexplored. Given that FL can aggregate models using diverse strategies, the chosen aggregation method can significantly influence training effectiveness.

To address the challenges associated with wind turbine blade icing prediction and drawing inspiration from the advantages of federated learning, we propose a novel federated learning approach called FedBIP. This approach incorporates both data and model enhancements to optimize prediction performance. From a data perspective, FedBIP employs feature selection to retain only the most informative features, thereby reducing the influence of redundant data that can burden training and degrade model quality. Furthermore, an over-sampling technique is applied to alleviate issues arising from imbalanced data distributions. On the model side, FedBIP integrates a sequential architecture combining Long Short-Term Memory (LSTM) networks for capturing temporal dependencies and a densely connected Convolutional Neural Network (CNN) for extracting latent features, thus enhancing feature extraction capabilities. During model aggregation, FedBIP diverges from conventional federated learning techniques that simply average client parameters. Instead, it incorporates a weighted aggregation method that considers data volume, timestamp alignment, and offset variations across clients. This strategy addresses complications introduced by inactive clients or those with smaller, less representative datasets. Moreover, FedBIP employs a knowledge distillation (KD) mechanism in the local training phase, where a teacher model is used to accelerate model convergence at the client level. This mechanism not only improves individual client model performance but also expedites the overall training process, ultimately enhancing the global model's accuracy and efficiency. In this study, we emphasize the importance of data volume, temporal alignment, and offset consistency across participating clients to mitigate challenges associated with inactive or undertrained clients possessing datasets of varying sizes.

The contributions of this paper are summarized as follows:

1. This paper introduces FedBIP, a federated learning (FL) model specifically designed for predicting icing on wind turbine blades. The approach begins with feature selection to identify informative features, followed by data resampling to address data imbalance. Subsequently, a sequential structure is employed to extract critical information. During each communication round, an aggregation method is used that accounts for data size, timestamp, and offset relevance of each participating client, enabling more effective parameter aggregation. Additionally, a knowledge distillation (KD) strategy is integrated into local client training to accelerate model convergence and enhance overall training efficiency.
2. FedBIP is evaluated using a real-world dataset from two wind farms, with experimental comparisons against other methods. The results demonstrate the superior performance of FedBIP. An ablation study is also conducted to validate the effectiveness of the model's components.

The remainder of this paper is organized as follows: Section 9.2 details the proposed FL-based model and methodology. Section 9.3 presents experimental results that demonstrate the model's effectiveness. Finally, Sect. 9.4 concludes the paper.

## 9.2 Methodology

In this section, we first present an overview of the proposed framework, followed by a detailed description of each constituent procedure.

### *9.2.1 Overview*

Applying Federated Learning (FL) to wind turbine icing prediction poses several challenges. Firstly, wind turbines are geographically dispersed, ranging from mountainous terrains to offshore environments, leading to significant variations in environmental conditions. Although each turbine is equipped with hundreds of sensors monitoring its operational status, only a subset of these sensors provides meaningful data for detecting blade icing. Therefore, a robust feature selection process is crucial to identify the relevant features for effective model training. Secondly, icing events are sporadic, primarily occurring during cold seasons like winter, resulting in dataset imbalance. As turbines typically operate without icing, the majority of collected data represents non-icing conditions. Training on such imbalanced datasets can bias the

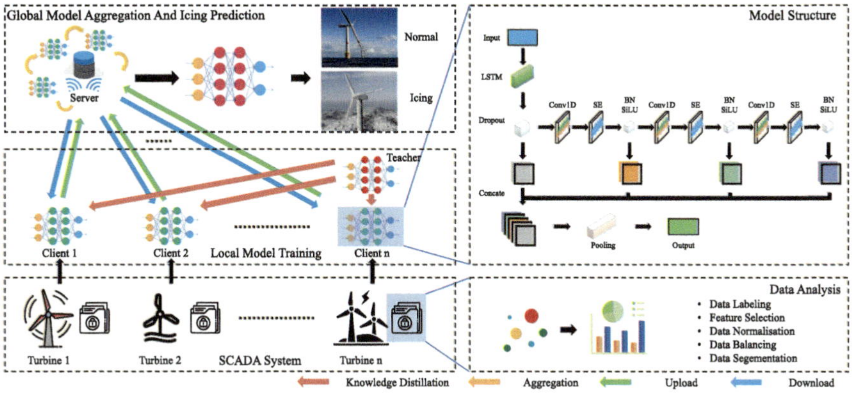


**Fig. 9.1** FL-based wind turbine icing prediction structure

model towards the majority class, reducing its accuracy in detecting icing. Finally, due to commercial privacy concerns, each turbine operates as an independent client within the FL framework. Managing variations in training data across clients is essential for developing a globally optimal model that can be trained efficiently.

To address these challenges, we propose a comprehensive icing prediction framework comprising three primary components, as illustrated in Fig. 9.1: (1) **Data Analysis**—Raw data from the Supervisory Control and Data Acquisition (SCADA) system undergoes a systematic processing pipeline including labeling, feature selection, normalization, data balancing, and segmentation. (2) **Local Model Training**—Processed data is distributed to individual clients, where each client trains a local model using its own data. (3) **Server Model Aggregation and Testing**—In each communication round, participating clients upload their locally trained model weights to a central server. The server then aggregates these weights using a weight aggregation algorithm, resulting in a final global model capable of predicting icing events. This structured approach aims to address the specific challenges of FL in wind turbine icing prediction, enabling robust model development across heterogeneous client data.

### *9.2.2 Data Analysis*

Modern wind turbines are equipped with hundreds of embedded sensors that continuously monitor diverse operational parameters. However, not all sensor data is valuable for model development, necessitating careful selection of relevant features. Furthermore, normal operating conditions are predominant, leading to significantly imbalanced datasets where abnormal data is sparse. This imbalance poses a challenge to model accuracy and sensitivity. Therefore, comprehensive data analysis, including relevance assessment and data balancing techniques, is essential to enhance model performance and sensitivity towards rare, abnormal events.

#### 9.2.2.1 Feature Selection

High-dimensional data significantly increases memory storage requirements and computational costs during data analysis. Moreover, learning models trained on extensive feature sets are prone to overfitting, often leading to diminished performance [34]. Wind turbine sensors collect numerous data attributes that vary considerably depending on the operating region and surrounding climate conditions. Even turbines in close proximity may generate data with distinct distributions. Furthermore, some attributes might be critical for detecting blade icing in one turbine but ineffective in others. Given our objective to develop a model using common attributes as input, a rigorous feature selection process is essential to identify the most relevant features for model training.

In this study, we employ the chi-square statistical test to evaluate the relevance score between sensor-derived columns and their corresponding labels. The equations for computing the chi-square score, $Schi_j$, for feature $k$, are

$$Schi_j = \sum_{k=1}^{2} \frac{(A_{kj} - B_{kj})^2}{B_{kj}}, \tag{9.1}$$

$$A = Y \cdot X, \tag{9.2}$$

$$B = Y' \cdot X'. \tag{9.3}$$

Let $Y \in \mathbb{R}^{2\times n}$ and $X \in \mathbb{R}^{n\times n_f}$, where $n$ is the number of input data samples and $n_f$ is the number of input features. The second row of the binary matrix $Y$ contains the ground-truth labels: 0 for normal instances and 1 for icing instances. The first row of $Y$ is complementary to the second row, with values of 1 minus the corresponding values in the second row. The matrix $X$ represents the input data, where each column corresponds to a specific feature across all samples. Let $Y' \in \mathbb{R}^{2\times 1}$, where each entry is the mean value of the corresponding row in $Y$, and $X' \in \mathbb{R}^{1\times n_f}$, where each element is the sum of all data points within the respective feature column of $X$.

For feature selection, we initially ranked the scores of each column for each wind turbine, producing a distinct rank list per turbine. The turbines in this study are located in two separate wind farms, where data distributions can vary, leading to different feature importances across farms. Even turbines within the same farm exhibit variability in relevant columns. To address this, we first selected columns common to each ranking list. As this subset proved insufficient for training, we augmented it by manually selecting high-ranking columns from each farm, prioritizing those highly ranked on one farm but less so on the other. Throughout this process, only the names and scores of each column were considered.

#### 9.2.2.2 Normalization

To mitigate the influence of individual data samples and reduce disparities in feature magnitudes across dimensions, the raw data undergoes a normalization procedure. This process transforms the numerical range of the data to the interval [0,1], ensuring consistency and comparability across features.

#### 9.2.2.3 Data Balancing

Wind turbines typically operate continuously, with intermittent shutdowns for faults like icing. Consequently, normal operation data significantly outnumber abnormal instances, resulting in highly imbalanced training datasets that negatively affect model performance, especially in predicting minority classes [35]. To mitigate the impact of data imbalance in training, resampling strategies are employed. Data balancing approaches broadly fall into two categories: generating synthetic data using mathematical methods, and over-sampling existing abnormal instances to achieve an approximately balanced dataset. This study focuses on the latter, utilizing available abnormal data to create additional abnormal samples.

During data segmentation, turbine data are framed within fixed-length time windows. No over-sampling is applied to normal operation data. For icing instances, however, a resampling strategy is implemented by advancing the time window with a fixed stride to generate more icing samples. Figure 9.2 illustrates the sampling process for both normal and icing data. The primary advantage of this approach is that it avoids synthetic data generation in both training and testing, preserving experimental integrity and dataset fidelity.

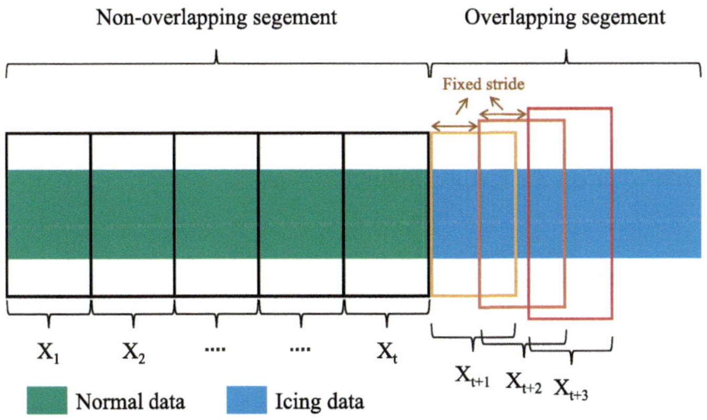


**Fig. 9.2** An illustration of data sample generation for normal samples and icing samples

Resampling is applied exclusively to training data, ensuring the test set remains untouched to maintain the authenticity of model evaluation. Following data balancing, segmentation is performed using a fixed-size window to accommodate the temporal continuity of raw data collected from wind turbine SCADA systems.

## 9.2.3 *Federated Learning*

### 9.2.3.1 Client and Server Model Design

In this paper, the client model and server model share an identical structure. The model developed for this study is a Long Short-Term Memory (LSTM) network combined with a three-layer densely connected Convolutional Neural Network (CNN) featuring Squeeze-and-Excitation (SE) [36] channel attention, as depicted in Fig. 9.1. For time-series sensor data from SCADA systems, we construct a model with two interconnected components: LSTM and CNN. LSTM is used to learn temporal features, which are then fed into a three-layer densely connected CNN to extract abstract features. For an input $X_{in} \in \mathbb{R}^{C \times TW}$ to the CNN, where $C$ is the number of features and $TW$ is the time window size, the output of each CNN layer component can be formulated as

$$X_{conv} = Conv1d(X_{in}), \tag{9.4}$$

$$X_{se} = SE(X_{conv}), \tag{9.5}$$

$$X_{norm} = Norm1d(X_{se}), \tag{9.6}$$

$$X_{out} = SiLU(X_{norm}), \tag{9.7}$$

where $X_{conv}$, $X_{se}$, $X_{norm}$, and $X_{out}$ are the outputs of Conv1d, SE, Norm1d, and SiLU, respectively. These variables share the same shape in each layer, $R^{H \times TW}$, where $H$ is the number of filters in the CNN. The three CNN layers are in series, with outputs $X_1$, $X_2$, and $X_3$. These outputs, along with the LSTM layer output $X_{lstm}$, are concatenated along the feature columns.

### 9.2.3.2 Aggregation of Global Model in FL

In federated learning, clients with larger datasets tend to train more effectively than those with smaller datasets. Treating all clients equally, regardless of data volume, can lead to suboptimal training, as clients with limited data may negatively impact overall model performance. Therefore, weighting clients based on data volume is beneficial for enhancing training robustness. Additionally, in each communication round, a subset of clients is randomly selected. Some clients may remain idle for extended periods, leading to less effective model updates compared to recently active clients. This disparity can hinder training, as models from less active clients may lag in accuracy and disrupt convergence when reintroduced [37]. Thus, the server should consider the time elapsed since each client's last participation to maintain consistency and model fidelity. Finally, client updates in each communication round can vary significantly. Aggregating these updates by simple summation may overemphasize parameters with higher absolute values, potentially skewing the global model. To prevent this, the server should prioritize alignment with the majority direction of client updates, ensuring balanced model improvement.

The central server aggregation process is as follows: In each communication round $c$, $K$ client models are randomly selected from all clients, where $K = pr \cdot N$, $pr$ is the participation rate and $N$ is the total number of clients. Each participating client trains its local model with $N_i$ samples, where $i$ is the $i$-th participating client's index. Initially, the server and all client models are randomly initialized.

For a client with learning rate $\epsilon$ and local gradient $g_i$, the local client update, $W_c^i$, is

$$W_{c-1}^i - \epsilon \cdot g_i \longrightarrow W_c^i. \tag{9.8}$$

After each participating client computes and applies its local gradient, the server aggregates weights from all participating client models to create a new global model, $W_c$. The aggregation process is mathematically formulated as

$$sign(x) = \begin{cases} 1, \text{if } x > 0 \\ 0, \text{if } x = 0 \\ -1, \text{if } x < 0 \end{cases} \tag{9.9}$$

$$\eta \cdot sign\left(\sum_{i=1}^{K} \frac{N_i}{N} \cdot \left(\frac{e}{2}\right)^{-(c-ts_i)} \cdot sign(W_c^i)\right) \rightarrow W_c, \tag{9.10}$$

here, $c$ is the current communication round index, $ts_i$ is the index of the last round client $i$ participated in, and $e$ is the base of the natural logarithm, modeling the temporal effect, controlled by hyperparameter $\eta$. $\frac{N_i}{N}$ weights data size, giving more weight to clients with larger datasets. $\left(\frac{e}{2}\right)^{-(c-ts_i)}$ is the time-based weight, favoring recently participating clients. Finally, $\text{sign}(W_c^i)$ guides the global model update, aligning with the majority update direction.

#### 9.2.3.3 Knowledge Distillation

"Cumbersome networks" are models requiring extensive training time to reach a global or satisfactory local optimum, demanding significant computational resources [38]. In classification models, the output represents class likelihood. The highest likelihood does not definitively confirm class membership but indicates close resemblance. This knowledge can transfer insights from a teacher to a student model.

Federated Learning (FL) distributes model training to collaborating clients, decentralizing server training. Ignoring client sample size, local client training resembles centralized training. Though client datasets are smaller, training from scratch is time intensive, requiring local optimum convergence. Models with many parameters can have prohibitively large model updates, needing numerous communication rounds for a high-quality global model [39]. Knowledge distillation can expedite training by reducing convergence time.

In this paper, the KD procedure is shown in Fig. 9.3. By introducing a teacher model, we calculate the Kullback–Leibler divergence loss ($L_{KL}$), then combine it with the cross-entropy loss ($L_{CE}$) to get the knowledge distillation loss ($L_{KD}$). The equations are

$$L_{KL} = KL\left(log\left(\frac{p_s}{T}\right) || \frac{p_t}{T}\right) = log\left(\frac{p_s}{T}\right) \cdot \frac{log\left(\frac{p_s}{T}\right)}{\frac{p_t}{T}}, \tag{9.11}$$

$$L_{CE} = -y \cdot log(p_s) + (1 - y) \cdot log(1 - p_s), \tag{9.12}$$

$$L_{KD} = \alpha \cdot T^2 \cdot L_{KL} + (1 - \alpha) \cdot L_{CE}, \tag{9.13}$$

where $p_s$ and $p_t$ are the predicted values of the student and teacher models, $y$ is the true value, $\alpha$ is a hyperparameter, and $T$ is the temperature for distilling "soft targets".

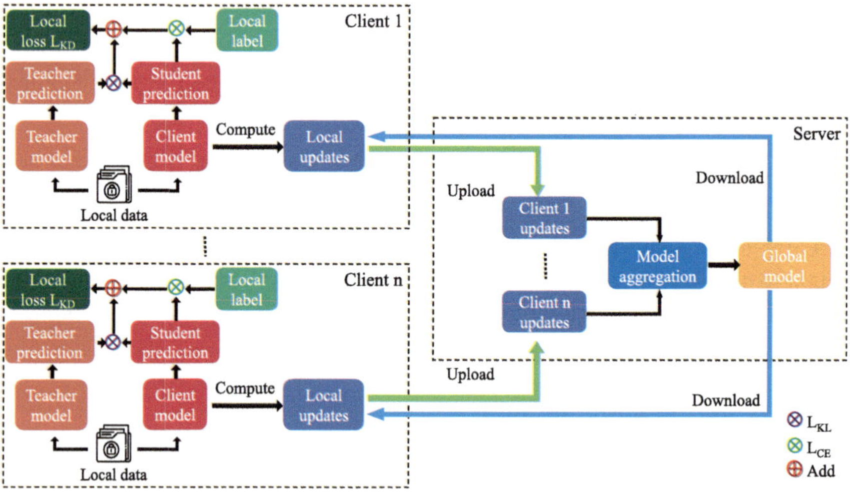


**Fig. 9.3** Procedure of distributed training with knowledge distillation

## 9.3 Experiment

### *9.3.1 Dataset*

This study uses a wind turbine dataset from two distinct wind farms about 600 kilometers apart: Yanling in Henan Province and Hongshimao in Shaanxi Province. The dataset contains raw data recorded at 30-second intervals over roughly 15 days, from February 11 to 26, 2019. The dataset has an imbalanced distribution of icing events, with Yanling experiencing 2 days and Hongshimao 1 day of icing. This results in data ratios of approximately 7:1 for Yanling and 15:1 for Hongshimao. Labels were assigned in consultation with experienced engineers. The dataset includes about 300 sensor-derived columns of string, float, and Boolean types. Table 9.1 lists the selected columns used in the analysis.

**Table 9.1** Specification of selected parameters

| No. | Parameter | No. | Parameter |
|---|---|---|---|
| 1 | Generator speed | 18 | Torque of generator |
| 2 | Speed of pitch1 | 19 | Temperature of shaft cabinet1 |
| 3 | Speed of pitch2 | 20 | Temperature of shaft cabinet2 |
| 4 | Angle of pitch1 | 21 | Temperature of shaft cabinet3 |
| 5 | Angle of pitch2 | 22 | Temperature of stator of generator U1 |
| 6 | Angle of pitch3 | 23 | Temperature of stator of generator U2 |
| 7 | Yawing angle | 24 | Temperature of stator of generator V1 |
| 8 | Temperature of hub1 | 25 | Temperature of stator of generator V2 |
| 9 | Temperature of hub2 | 26 | Temperature of stator of generator W1 |
| 10 | Temperature of hub3 | 27 | Temperature of stator of generator W2 |
| 11 | Power performance1 | 28 | 60s mean of wind speed |
| 12 | Power performance2 | 29 | Current1 |
| 13 | Power performance3 | 30 | Current1 |
| 14 | Temperature of pitch motor1 | 31 | Current1 |
| 15 | Temperature of pitch motor2 | 32 | Temperature of cooling pad1 |
| 16 | Temperature of pitch motor3 | 33 | Temperature of cooling pad2 |
| 17 | Active power | 34 | Temperature of cooling pad3 |

### 9.3.2 Experiment Settings

In data analysis, the time window size ($TW$) is set to 128, with a stride of 70 for minority class over-sampling. Segmentation shifts the time window three times to augment the minority class sample set. Training involves 20 clients, each with data from a distinct turbine. The participation rate is 0.5, meaning 10 turbines are randomly selected per communication round. Adam optimizer is used with a learning rate of 0.01. Over 100 communication rounds, a central server is trained to deliver a global model with optimal performance. The local client update learning rate ($\epsilon$) is 0.01 and the model aggregation hyperparameter ($\eta$) is 0.04. For knowledge distillation (KD), the weighting factor ($\alpha$) is 0.9 and the temperature parameter ($T$) is 20.

### 9.3.3 Evaluation Metrics

To evaluate performance, we use $F_\beta$_Score and Balanced Accuracy (BA). Given the imbalanced dataset (8:1 imbalance ratio) and the focus on icing prediction sensitivity, conventional accuracy is inappropriate. Predicting all samples as normal could yield high accuracy, but not demonstrate model effectiveness. BA is thus a more reliable metric. The equations are

$$F_\beta = (1 + \beta^2)\frac{Precision \cdot Recall}{\beta^2 \times Precision + Recall}, \tag{9.14}$$

$$Precision = \frac{TP}{TP + FP}, \quad Recall = \frac{TP}{TP + FN}, \tag{9.15}$$

where TP, FP, TN, and FN are True Positive, False Positive, True Negative, and False Negative, respectively. $\beta$ is set to 2.

### 9.3.4 Data Balancing

We compare our sliding window data balancing method with SMOTE [40], an over-sampling method that creates synthetic examples by interpolating minority class samples with their $k$ nearest neighbors ($k = 5$). Model performance comparisons are in Fig. 9.4.

Both SMOTE and Sliding Window improve performance over the imbalanced dataset. However, SMOTE generates many synthetic samples, potentially compromising experimental authenticity and failing to capture temporal dependencies. Sliding Window effectively uses existing icing data, leading to better model performance

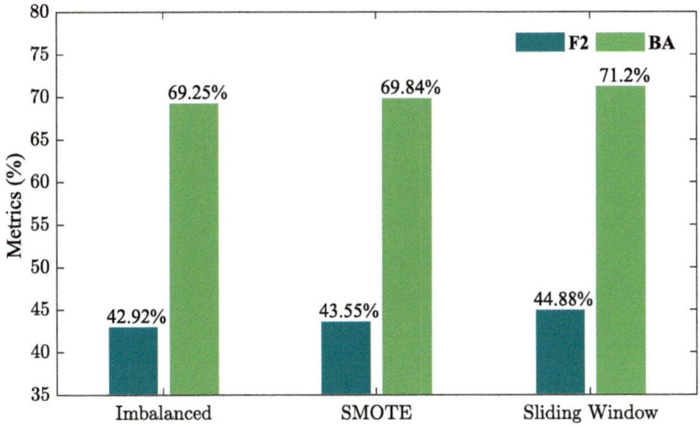


**Fig. 9.4** Performance comparison of over-sampling methods

without overfitting synthetic data. Over-representation of synthetic icing samples could negatively affect model generalization.

### 9.3.5 *Model Evaluation*

We compare FedBIP against seven benchmark neural networks for time-series classification:

1. **BiLSTM**: Bidirectional LSTM with forward and backward LSTM layers, each with 128 hidden units.
2. **CNN-LSTM**: Hybrid model with a CNN layer followed by an LSTM layer, both with 128 filters.
3. **SSENET**: Robust baseline for multivariate time-series classification with dense connections. Attention mechanism excluded for fair structural comparison.
4. **FCN**: Fully convolutional network for time-series classification [41]. Three convolutional layers with 128, 256, and 128 filters, respectively.
5. **GRU**: Gated recurrent unit, a computationally efficient LSTM variant, with 128 hidden units.
6. **LSTM**: Widely used RNN for time-series classification, with 128 hidden units.
7. **MLP**: Multilayer perceptron with three hidden layers, each with 128 units. Dropout layers between hidden layers to mitigate overfitting.

Table 9.2 shows FedBIP outperforms other models on both imbalanced and over-sampled datasets. On the over-sampled dataset, FedBIP improves $F_2$ by 1.68% and BA by 0.95% over FCN. On the imbalanced dataset, FedBIP improves $F_2$ by 0.54% and BA by 0.51% over SSENET. Most models improve on the over-sampled dataset, except SSENET. Despite complexity and longer training, SSENET ranks

**Table 9.2** Model performance comparison on imbalanced dataset and over-sampled by sliding window dataset

| | Imbalanced | | Over-sampled | |
|---|---|---|---|---|
| | $F_2$ (%) | BA (%) | $F_2$ (%) | BA (%) |
| BILSTM | 42.36 | 68.68 | 42.52 | 68.81 |
| CNN_LSTM | 40.41 | 66.15 | 41.67 | 67.94 |
| SSENET | 42.69 | 68.90 | 40.71 | 66.80 |
| FCN | 42.61 | 68.84 | 44.14 | 70.53 |
| GRU | 41.79 | 68.07 | 42.18 | 68.48 |
| LSTM | 41.64 | 67.88 | 42.90 | 69.27 |
| MLP | 42.41 | 68.78 | 43.36 | 69.83 |
| FedBIP | **42.92** | **69.25** | **44.88** | **71.20** |

second on original data but performs poorest on over-sampled data. Model structure choice is crucial; complexity does not guarantee better performance and can lead to degradation with increased computational cost and training time.

### *9.3.6 Model Aggregation Comparison*

We compare our aggregation method against:

1. **Federated Averaging (FedAVG)** [25]: Baseline, weighted average of parameters based on client sample size.
2. **Asynchronous Temporally Weighted Aggregation (ASTW)** [37]: FedAVG variant with temporal information. We focus on the temporally weighted (TW) module.
3. **Majority Voting (MV)**: Aggregated parameters determined by the sign (positive/negative) of each parameter across most clients.

Figure 9.5 shows our method outperforms others. Compared to MV, our method improves $F_2$ by 1.88% and BA by 1.09%. Relative to TW, improvements are 3.27% in $F_2$ and 2.4% in BA. FedAVG performs poorest, with our method surpassing it by 4.28% in $F_2$ and 2.68% in BA.

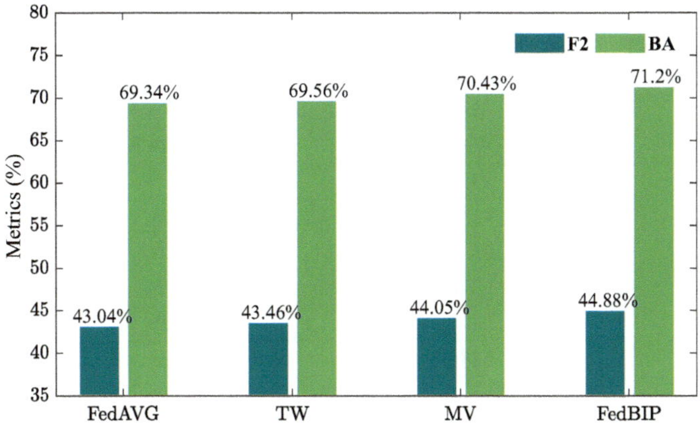


**Fig. 9.5** Model aggregation comparison on over-sampled by sliding window dataset

### 9.3.7 Model with Knowledge Distillation Evaluation

We assess Knowledge Distillation (KD) efficacy by comparing models with and without KD. An optimal model trained from scratch on the over-sampled dataset is used as the teacher model. Results are in Table 9.3.

KD generally improves performance across most models on both datasets. On the imbalanced dataset, KD makes performance resemble the teacher model. On the over-sampled dataset, FCN and GRU performance slightly decreased after KD, possibly because the teacher model mitigates imbalance. However, these models still outperform those without KD. CNN_LSTM and SSENET significantly improve with KD, outperforming the teacher model. CNN_LSTM improves by 7.86% $F_2$ and 4.71% BA, ranking third. SSENET improves and ranks second. Even training the

**Table 9.3** Performance comparison of model training with knowledge distillation on imbalanced and over-sampled by sliding window dataset

| | Imbalanced | | Over-sampled | |
|---|---|---|---|---|
| | $F_2$ (%) | BA (%) | $F_2$ (%) | BA (%) |
| BILSTM | 43.30 | 69.64 | 44.00 | 70.16 |
| CNN_LSTM | 43.39 | 69.70 | 44.97 | 71.14 |
| SSENET | 42.89 | 69.29 | 45.42 | 71.57 |
| FCN | **44.70** | **70.97** | 43.72 | 69.89 |
| GRU | 43.16 | 69.49 | 40.03 | 66.34 |
| LSTM | 44.10 | 70.30 | 43.68 | 69.82 |
| MLP | 43.78 | 69.93 | 43.41 | 69.75 |
| FedBIP | 44.41 | 70.55 | **45.96** | **71.45** |

**Table 9.4** The average number of communication rounds to reach the target BA of 63% of 10 times with or without KD on the over-sampled by sliding window dataset (The fewer communication rounds, the fewer communication costs)

| | Average communication rounds (10 runs) | |
|---|---|---|
| | Without KD | With KD |
| BILSTM | 48.2 | 32 |
| CNN_LSTM | 41.8 | 26.6 |
| SSENET | 63.4 | 36.6 |
| FCN | 50.6 | 47.5 |
| GRU | 54 | 48.5 |
| LSTM | 64.6 | 18.6 |
| MLP | 51.1 | 45.3 |
| FedBIP | 31 | 29.4 |

teacher model structure from scratch with KD on the over-sampled dataset improves performance by 2.41% $F_2$ and 0.35% BA.

We compare communication rounds to reach 63% BA with and without KD. Table 9.4 shows average rounds over 10 runs for each model.

All models require fewer rounds with KD. LSTM shows the largest improvement, reducing rounds by 71.21%. SSENET reduces rounds by 42.27%. FedBIP shows a modest 5.16% reduction, but KD still accelerates training.

## *9.3.8 Ablation Analysis*

Two sensitivity analyses: (1) FedBIP with different modules; (2) Data balancing with different imbalance ratios.

### 9.3.8.1 Impact of FedBIP with Different Modules

Ablation experiments compare FedBIP with: (1) FedBIP_ReLU (ReLU activation); (2) FedBIP_noSE (no SE blocks); (3) FedBIP_Focal (Focal Loss, $\gamma = 4, \alpha = 0.25$). Results in Fig. 9.6.

Focal loss causes the largest performance drop (4.66% $F_2$, 2.84% BA). Removing SE blocks drops performance by 4.15% $F_2$ and 2.58% BA. ReLU activation causes marginal degradation (1.33% $F_2$, 0.92% BA).

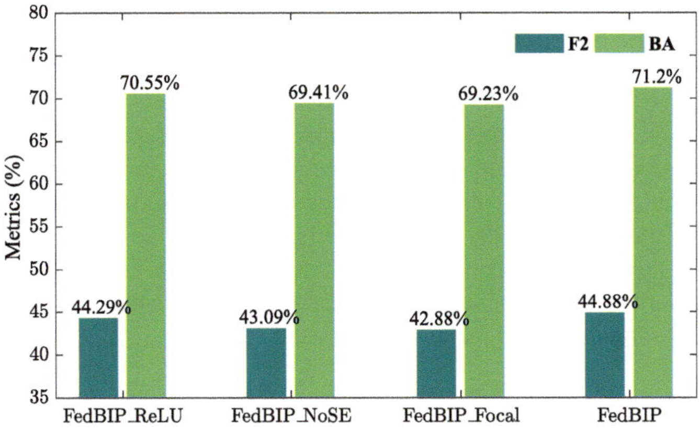


**Fig. 9.6** Model ablation experiment on over-sampled by sliding window dataset

#### 9.3.8.2 Impact of Data Balancing with Different Imbalance Ratios

We investigate the impact of sliding window over-sampling with varying imbalance ratios. Each sample generation iteration produces icing samples equal to the number of icing instances in the imbalanced dataset. Results in Fig. 9.7.

Performance improves as the dataset becomes more balanced. Optimal performance is at imbalance ratios of 19:5 (Hongshimao) and 9:5 (Yanling). Further increasing icing samples decreases performance due to overfitting minority class samples, negatively impacting generalization and icing prediction accuracy.

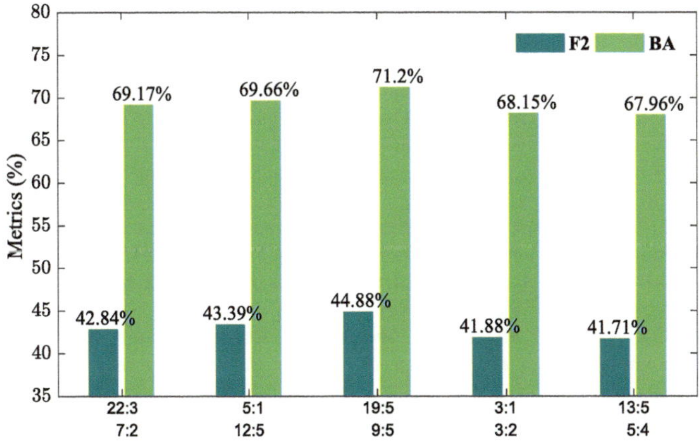


**Fig. 9.7** Performance comparison on over-sampled by sliding window dataset with different imbalance ratios (the first row is the imbalance ratio of data of Hongshimao, and the second row is the imbalance ratio of data of Yanling)

## 9.4 Conclusion

This paper introduces FedBIP for wind turbine blade icing prediction, incorporating feature selection, federated learning, and knowledge distillation. FedBIP trains local models on client data and aggregates them centrally without raw data sharing. A weighted aggregation method and knowledge distillation enhance model performance. Experiments on data from 20 turbines across two farms show FedBIP outperforms other approaches.

Future work will address class imbalance algorithmically and explore inter-column data correlations for feature selection.

## References


1. Veers P, Dykes K, Lantz E, Barth S, Bottasso CL, Carlson O, Clifton A, Green J, Green P, Holttinen H et al (2019) Grand challenges in the science of wind energy. Science 366(6464):eaau2027
2. Sadorsky P (2021) Wind energy for sustainable development: driving factors and future outlook. J Cleaner Prod 289:125779
3. Wei K, Yang Y, Zuo H, Zhong D (2020) A review on ice detection technology and ice elimination technology for wind turbine. Wind Energy 23(3):433–457
4. Wang P, Zhou W, Bao Y, Li H (2018) Ice monitoring of a full-scale wind turbine blade using ultrasonic guided waves under varying temperature conditions. Struct Control Health Monitoring 25(4):e2138
5. Xu B, Lu F, Dyke SJ, Guo X (2014) Icing monitoring for a wind turbine model blade with active pzt technology. Earth and Space 2014:703–710
6. Bin X, Xueyi G, Hongbing C (2017) Active icing monitoring for wind turbine blade models with pzt technology. Piezoelectrics & Acoustooptics 39:02
7. Zhao X, Rose JL (2016) Ultrasonic guided wave tomography for ice detection. Ultrasonics 67:212–219
8. Lega M, Ferrara C, Persechino G, Bishop P (2014) Remote sensing in environmental police investigations: aerial platforms and an innovative application of thermography to detect several illegal activities. Environ Monitoring Assess 186(12):8291–8301
9. Musto M, Rotondo G, De Cesare M, Del Vecchio A, Savino L, De Filippis F (2016) Error analysis on measurement temperature by means dual-color thermography technique. Measurement 90:265–277
10. Venna SV, Lin Y-J, Botura G (2007) Piezoelectric transducer actuated leading edge de-icing with simultaneous shear and impulse forces. J Aircraft 44(2):509–515
11. Szasz R-Z, Ronnfors M, Revstedt J (2016) Influence of ice accretion on the noise generated by an airfoil section. Int J Heat Fluid Flow 62:83–92
12. Skrimpas GA, Kleani K, Mijatovic N, Sweeney CW, Jensen BB, Holboell J (2016) Detection of icing on wind turbine blades by means of vibration and power curve analysis. Wind Energy 19(10):1819–1832
13. Roberge P, Lemay J, Ruel J, Bégin-Drolet A (2018) A new atmospheric icing detector based on thermally heated cylindrical probes for wind turbine applications. Cold Regions Sci Technol 148:131–141
14. Björkman M (2004) Long time measurements of noise from wind turbines. J Sound Vib 277(3):567–572
15. Pedersen MC, Yin C (2014) Preliminary modelling study of ice accretion on wind turbines. Energy Procedia 61:258–261

16. Swenson L, Gao L, Hong J, Shen L (2022) An efficacious model for predicting icing-induced energy loss for wind turbines. Appl Energy 305:117809
17. Shu L, Qiu G, Hu Q, Jiang X, McClure G, Yang H (2018) Numerical and field experimental investigation of wind turbine dynamic de-icing process. J Wind Eng Indus Aerodyn 175:90–99
18. Chen L, Xu G, Zhang Q, Zhang X (2019) Learning deep representation of imbalanced scada data for fault detection of wind turbines. Measurement 139:370–379
19. Yuan B, Wang C, Jiang F, Long M, Philip SY, Liu Y (2019) Waveletfcnn: a deep time series classification model for wind turbine blade icing detection
20. Tian W, Cheng X, Li G, Shi F, Chen S, Zhang H (2021) A multilevel convolutional recurrent neural network for blade icing detection of wind turbine. IEEE Sensors J 21(18):20 311–20 323
21. Tong R, Li P, Lang X, Liang J, Cao M (2021) A novel adaptive weighted kernel extreme learning machine algorithm and its application in wind turbine blade icing fault detection. Measurement 185:110009
22. Cheng X, Shi F, Zhao M, Li G, Zhang H, Chen S (2021) Temporal attention convolutional neural network for estimation of icing probability on wind turbine blades. IEEE Trans Indus Electron
23. Wang X, Zheng Z, Jiang G, He Q, Xie P (2022) Detecting wind turbine blade icing with a multiscale long short-term memory network. Energies 15(8):2864
24. Xiao J, Li C, Liu B, Huang J, Xie L (2022) Prediction of wind turbine blade icing fault based on selective deep ensemble model. Knowl Based Syst 242:108290
25. McMahan B, Moore E, Ramage D, Hampson S, y Arcas BA (2017) Communication-efficient learning of deep networks from decentralized data. In: Artificial intelligence and statistics. PMLR, pp 1273–1282
26. Hard A, Rao K, Mathews R, Ramaswamy S, Beaufays F, Augenstein S, Eichner H, Kiddon C, Ramage D (2018) Federated learning for mobile keyboard prediction. arXiv preprint arXiv:1811.03604
27. Voigt P, Von dem Bussche A (2017) The eu general data protection regulation (gdpr). A Practical Guide, 1st Ed., Cham: Springer International Publishing, vol 10, no 3152676, pp 10–5555
28. Sheller MJ, Edwards B, Reina GA, Martin J, Pati S, Kotrotsou A, Milchenko M, Xu W, Marcus D, Colen RR et al (2020) Federated learning in medicine: facilitating multi-institutional collaborations without sharing patient data. Scientific Reports 10(1):1–12
29. Wang Y, Bennani IL, Liu X, Sun M, Zhou Y (2021) Electricity consumer characteristics identification: a federated learning approach. IEEE Trans Smart Grid 12(4):3637–3647
30. Cheng X, Tian W, Shi F, Zhao M, Chen S, Wang H (2022) A blockchain-empowered cluster-based federated learning model for blade icing estimation on iot-enabled wind turbine. IEEE Trans Indus Inform
31. Pei J, Zhong K, Jan MA, Li J (2022) Personalized federated learning framework for network traffic anomaly detection. Comput Networks 209:108906
32. Cheng X, Shi F, Liu Y, Zhou J, Liu X, Huang L (2022) A class-imbalanced heterogeneous federated learning model for detecting icing on wind turbine blades. IEEE Trans Indus Inform
33. Arikumar K, Prathiba SB, Alazab M, Gadekallu TR, Pandya S, Khan JM, Moorthy RS (2022) Fl-pmi: federated learning-based person movement identification through wearable devices in smart healthcare systems. Sensors 22(4):1377
34. Li J, Cheng K, Wang S, Morstatter F, Trevino RP, Tang J, Liu H (2017) Feature selection: a data perspective. ACM Comput Surv (CSUR) 50(6):1–45
35. Wang L, Xu S, Wang X, Zhu Q (2020) Towards class imbalance in federated learning. arXiv e-prints, pp arXiv–2008
36. Hu J, Shen L, Sun G (2018) Squeeze-and-excitation networks. In: Proceedings of the IEEE conference on computer vision and pattern recognition, pp 7132–7141
37. Chen Y, Sun X, Jin Y (2019) Communication-efficient federated deep learning with layer-wise asynchronous model update and temporally weighted aggregation. IEEE Trans Neural Networks Learn Syst 31(10):4229–4238
38. Yim J, Joo D, Bae J, Kim J (2017) A gift from knowledge distillation: Fast optimization, network minimization and transfer learning. In: Proceedings of the IEEE conference on computer vision and pattern recognition, pp 4133–4141

39. Wu C, Wu F, Lyu L, Huang Y, Xie X (2022) Communication-efficient federated learning via knowledge distillation. Nature Commun 13(1):1–8
40. Chawla NV, Bowyer KW, Hall LO, Kegelmeyer WP (2002) Smote: synthetic minority over-sampling technique. J Artif Intell Res 16:321–357
41. Wang Z, Yan W, Oates T (2017) Time series classification from scratch with deep neural networks: a strong baseline. In: International joint conference on neural networks (IJCNN). IEEE, pp 1578–1585

# Chapter 10
# Blockchain-Empowered Clustered Federated Convolutional Neural Network for Blade Icing Detection

## 10.1 Introduction

Wind energy has solidified its position as a vital and rapidly growing renewable energy resource, underpinned by its extensive availability and the progressive maturity of wind turbine technologies [1]. Despite its considerable promise, a major impediment to the efficiency and dependability of wind turbines is blade icing. This phenomenon can severely curtail turbine functionality, leading to significant reductions in annual power generation, potentially reaching up to 30% in extreme icing conditions [2]. Consequently, addressing blade icing is of utmost importance for ensuring the financial viability and operational effectiveness of wind farms, particularly in regions susceptible to cold climates.

To mitigate the adverse impacts of blade icing, a range of strategies have been devised, broadly categorized as passive and active methods. Passive anti-icing approaches commonly involve applying specialized coatings to blade surfaces [3]. These coatings aim to diminish ice adhesion and accumulation by modifying the surface characteristics. However, while coatings can offer some level of protection, they are often inadequate in completely preventing icing, especially under severe conditions, thus necessitating the consideration of active methodologies. Active de-icing and anti-icing techniques employ heating systems integrated into the blades to either prevent ice formation or facilitate its removal post-accretion [4]. Although generally more efficacious than passive methods, active systems often entail greater energy consumption and increased operational expenses. Critically, both passive and active methods depend on prompt and precise icing detection as a prerequisite for effective deployment. Without dependable detection systems, anti-icing systems may be activated unnecessarily, squandering energy and resources, or, conversely, icing events may go undetected, resulting in compromised performance and possible structural damage to turbine blades.

Traditional icing detection methodologies can be broadly classified based on several approaches: monitoring the physical attributes of ice, analyzing turbine behavior



X. Cheng et al., *Computational Methods for Blade Icing Detection of Wind Turbines*, Engineering Applications of Computational Methods 24,
https://doi.org/10.1007/978-981-96-6763-5_10

affected by ice buildup, and employing mathematical or numerical models. Methods based on physical properties often involve deploying specialized sensors to directly measure ice accumulation. Examples include optical sensing techniques, as developed by Shajiee et al. [5], and ultrasonic guided wave sensors, as investigated by Wang et al. [6]. Optical methods, while providing direct icing detection, can be vulnerable to environmental disturbances such as fog or snow, potentially leading to inaccurate readings. Ultrasonic methods, which track alterations in elastic wave propagation due to ice, necessitate sophisticated signal processing and careful sensor placement to ensure dependable detection amid operational noise and varying environmental conditions. Another category of conventional methods concentrates on indirectly detecting icing by analyzing shifts in turbine operational data. These methods monitor operational data from wind turbines to infer icing conditions based on deviations in performance. For instance, changes in rotor angular velocity, power output, or vibration patterns can indicate ice accumulation [7, 8]. While cost-effective as they utilize existing turbine sensors, these indirect methods may be less sensitive to the initial phases of icing and can be influenced by other operational variables like wind fluctuations or mechanical faults, potentially leading to false alarms or missed detections. Mathematical and numerical models offer a predictive approach to icing detection by simulating ice accretion based on meteorological data and blade geometry. However, the accuracy of these models is highly contingent on the quality of meteorological inputs, the complexity of the icing physics captured, and frequently requires substantial domain expertise for model development, calibration, and validation [9]. Overall, while traditional methods have significantly contributed to icing detection, they often encounter limitations concerning cost-effectiveness, precision in complex and fluctuating environmental conditions, and reliance on specific domain knowledge and dedicated sensor infrastructure.

In recent years, data-driven approaches, especially end-to-end solutions utilizing deep neural networks (DNNs), have emerged as a compelling alternative for icing detection, offering the potential to overcome some of the limitations of traditional methodologies. These methods excel at learning complex patterns directly from data, diminishing the need for explicit, manually engineered features or extensive domain-specific knowledge. While deep learning models circumvent the necessity for extensive prior knowledge, their efficacy is intrinsically linked to the availability of large, representative, and high-quality datasets. The rapid proliferation of Internet of Things (IoT) technologies within wind farms has resulted in a surge in the availability of operational data. Wind turbines are now equipped with numerous sensors that continuously stream data pertaining to power generation, temperature, vibration, and various other operational parameters. This abundance of IoT data presents a significant opportunity to leverage data-driven techniques for improved icing detection. By exploiting the rich spatio-temporal correlation information embedded within time-series data from multiple wind turbines and across different wind farms, the performance and robustness of icing detection models can be substantially enhanced. The prospect of improved detection accuracy and reliability through collaborative data utilization can incentivize greater engagement from wind farm operators in shared detection services and data-driven solutions [10].

However, realizing the complete potential of data-driven icing detection through the aggregation and analysis of IoT data from multiple wind farms encounters substantial challenges related to data privacy and the practicalities of centralized data processing. Firstly, wind farm owners are understandably hesitant to share sensitive raw operational data due to legitimate privacy concerns and commercial sensitivities. Detailed operational parameters are crucial for maintaining optimal turbine performance, refining maintenance schedules, and can be considered proprietary information that provides a competitive edge in the energy market. This reluctance to share raw data can significantly impede the feasibility of centralized data aggregation, which is the conventional approach for training potent machine learning models. Secondly, traditional machine learning paradigms are typically centralized, necessitating the consolidation of data from geographically dispersed sources into a central data repository. This centralized approach not only raises serious data privacy and security concerns but also becomes increasingly inefficient and computationally demanding as the volume of IoT data expands exponentially. The sheer magnitude of data generated by contemporary IoT-enabled wind turbines can overwhelm centralized processing infrastructure, making model training computationally expensive, time-consuming, and difficult to scale. Furthermore, managing and securing massive centralized datasets introduces considerable logistical and security complexities.

Federated learning (FL) has emerged as a persuasive distributed machine learning paradigm that directly addresses these challenges. FL provides a privacy-preserving methodology for collaborative machine learning, enabling the training of a global model across numerous decentralized clients (e.g., individual wind turbines or entire wind farms) without requiring the direct sharing or centralization of raw data [11]. Instead of exchanging raw data, FL clients locally train machine learning models on their respective datasets and only share model updates, such as gradients or model parameters, with a central server for aggregation. This approach considerably enhances data privacy by ensuring that sensitive raw operational data remains localized at each wind farm, mitigating the risks associated with data breaches and privacy violations.

Despite the inherent advantages of FL regarding privacy preservation and distributed computation, its direct application to icing detection in wind turbines still presents critical challenges. A primary concern is the reliance on a single centralized server in traditional FL frameworks. This central server acts as a single point of failure and a potential bottleneck in the FL network. The robustness, reliability, and trustworthiness of this central server are paramount, as its failure can disrupt the entire federated learning process. Moreover, the server is responsible for crucial functions such as selecting participating clients for each training round and aggregating model updates received from clients. These functions, if not implemented fairly and transparently, are susceptible to bias or manipulation, potentially compromising the fairness, robustness, and overall performance of the resultant global model. Another significant vulnerability of traditional FL is its susceptibility to various security threats, including malicious attacks. For instance, malicious clients could intentionally upload poisoned model updates to the server, aiming to degrade the performance of the aggregated global model or even introduce backdoors for adversarial purposes.

Furthermore, the central server itself, despite being intended as a trusted aggregator, could potentially be compromised or even be inherently malicious. In such scenarios, the privacy of all participating clients could be severely jeopardized, as the server has access to aggregated model updates which, under certain circumstances, can be exploited to infer private information about individual clients' datasets. Finally, a prevalent characteristic of wind turbine operational data in the context of icing detection is the inherent class imbalance. Wind turbines typically operate under normal, non-icing conditions for the vast majority of their operational lifespan. Icing events are relatively infrequent and tend to occur during specific periods characterized by low temperatures and high humidity. This natural phenomenon results in a highly skewed class distribution where normal operation samples significantly outnumber icing samples. Such imbalanced datasets pose a considerable challenge for training effective icing detection models. Standard classifiers trained on imbalanced data tend to be biased towards the majority class (normal operation) and often exhibit poor performance in detecting the minority class of interest (icing events), which is precisely what we aim to identify accurately.

To address the limitations of conventional centralized FL, particularly its reliance on a single point of failure and vulnerability to attacks, and to effectively handle the challenge of data imbalance inherent in icing detection datasets, this paper proposes a novel **BL**ockchain-empowered imb**A**lanced federate**D** l**E**arning model, termed BLADE. BLADE innovatively integrates blockchain technology into the federated learning framework to decentralize the model aggregation and update process, thereby enhancing robustness, security, and transparency by eliminating the single point of failure associated with a centralized server. Blockchain's intrinsic security features, including immutability, cryptographic integrity, and decentralized consensus mechanisms, further strengthen data privacy and ensure the integrity of model updates within the federated network. To specifically address the issue of imbalanced sensor data, BLADE incorporates a novel cluster-based imbalance learning method. This method is designed to improve the model's ability to effectively detect icing events, even when they are significantly underrepresented in the training data, by focusing on learning effective representations for the minority class. In summary, the key contributions of this paper are

1. We introduce **BLADE**, a **BL**ockchain-empowered imb**A**lanced federate**D** l**E**arning model, purpose-built for wind turbine blade icing detection. BLADE mitigates the inherent vulnerabilities of centralized servers in traditional FL systems by utilizing a decentralized and secure blockchain infrastructure for model aggregation and management. A novel cluster-based imbalance learning methodology is seamlessly integrated within the BLADE framework to adeptly manage the skewed class distributions that are characteristic of wind turbine sensor data in icing detection applications. BLADE significantly improves data privacy and ensures secure and auditable model updates within a trustless blockchain environment, making it particularly well suited for applications dealing with sensitive operational data and imbalanced event frequencies.

2. The effectiveness and resilience of the proposed BLADE model are rigorously evaluated using an extensive real-world dataset acquired from 10 wind turbines operating across two distinct wind farms. Comprehensive experimental results highlight BLADE's superior performance in blade icing detection when compared to conventional centralized models and standard federated learning approaches. Furthermore, we present detailed performance comparisons concentrating on global model accuracy, the efficacy of the imbalance learning capabilities, and the model's resilience against poisoning attacks to explicitly demonstrate the advantages and security benefits of BLADE. Thorough sensitivity analyses and ablation studies are conducted to systematically validate the effectiveness of each core component of the proposed model architecture and the integrated cluster-based imbalance learning strategy, providing insights into their individual contributions to the overall performance of BLADE.

The subsequent sections of this paper are organized as follows: Sect. 10.2 provides a comprehensive and detailed exposition of the proposed BLADE model, delineating its architecture, the integration of blockchain technology, and the cluster-based imbalance learning mechanism. Section 10.3 elaborates on the experimental design, presents the detailed outcomes of our evaluations, and offers thorough analyses of the findings to demonstrate the performance and advantages of BLADE. Finally, Sect. 10.4 concludes the paper by summarizing the principal contributions, emphasizing the key findings, and outlining promising avenues for future research and development in the domain of blockchain-empowered federated learning for wind turbine blade icing detection and related applications.

## 10.2 Methodology

### *10.2.1 Overview*

Blockchain can be conceptualized as an expanding chain of blocks, where individual blocks are interconnected using cryptographic methods [12, 13]. Blockchain enhances FL through several technical dimensions: (1) It provides a fully decentralized platform for round demarcation, client selection, and model aggregation. Leveraging distributed consensus mechanisms, blockchain can enhance the trustworthiness of FL processes. (2) Blockchain establishes a peer-to-peer network, improving the fault tolerance of the single server typically used in FL. This enhancement contributes to the computational flexibility and integrity of FL. (3) Blockchain can incentivize participant engagement by offering rewards for model sharing, facilitated through customizable smart contracts.

Despite the numerous advantages of integrating blockchain with FL, several challenges remain for deploying blockchain-based FL for blade icing detection. Firstly, addressing sensor data imbalance is crucial. Wind turbines predominantly operate in normal states, i.e., blades are ice free, with icing occurring only sporadically. Consequently, data from SCADA systems inherently exhibit imbalanced distributions, leading to biased detection models when trained on such data. Secondly, given the critical importance of wind turbines for wind farms and the broader transition to a low-carbon society, enhancing the security and reliability of blockchain-based FL for blade icing detection is paramount. It is essential to promote FL participation exclusively among legitimate entities while excluding malicious actors [14]. Moreover, providing adaptable consensus mechanism strategies is necessary, enabling users to select mechanisms based on their specific requirements. Thirdly, variations in the distribution and size of training data can lead to heterogeneity across different clients. For instance, wind turbines operate under diverse weather conditions and are managed by different operators. This heterogeneity poses a significant challenge to aggregating locally trained models into a robust global model for blade icing detection.

The *BLADE* model is designed to tackle these challenges, as depicted in Fig. 10.1. Unlike conventional blockchain systems, BLADE introduces a validation mechanism to bolster security. It distinguishes three roles: worker $\mathcal{W}$, validator $\mathcal{V}$, and miner $\mathcal{M}$, in contrast to the two roles (worker $\mathcal{W}$ and miner $\mathcal{M}$) in standard blockchains [14]. In BLADE, we consider $\mathcal{N}$ wind turbines with uniform computational capabilities. The operational flow of BLADE is as follows: **(1) Role Assignment**: Each wind turbine can assume one of three roles: worker $w_i \in \mathcal{W}$, validator $v_i \in \mathcal{V}$, or miner $m_i \in \mathcal{M}$, where the sum of these roles equals the total number of turbines, $\mathcal{W} + \mathcal{V} + \mathcal{M} = \mathcal{N}$. Role assignment in BLADE can be randomized or user defined. **(2) Local Imbalanced Learning**: Each worker $w_i$ performs imbalanced learning on its local dataset using a proposed cluster-based method, as illustrated in the top left of Fig. 10.1. **(3) Validation**: After local learning, each worker $w_i$ broadcasts its local model to an associated validator $v_i$. Validators then exchange their models with other validators and assess the legitimacy of local model updates. **(4) Mining**: Verified local

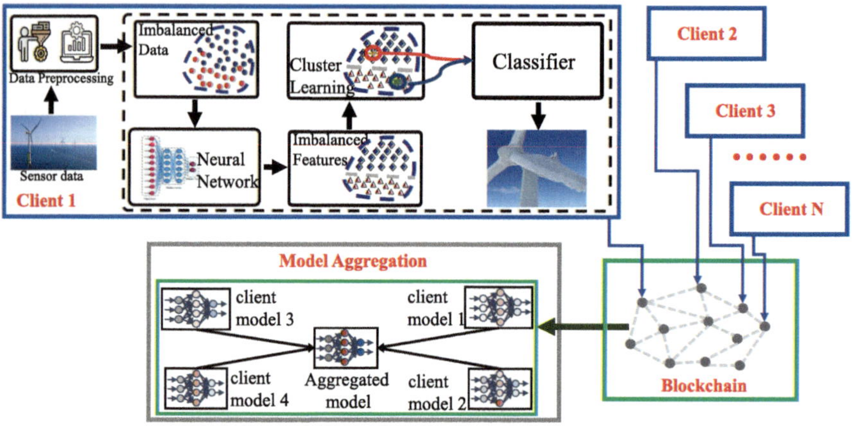


**Fig. 10.1** Architecture of blockchain-based federated learning systems for blade icing detection of wind turbine using IoT

models are transmitted to miners, who append a legitimacy block to their respective blockchains. Each miner requests its associated worker and validator to download this legitimacy block to their own blockchains. Once a new block is verified, the model within it becomes immutable. Without central server intervention, each client performs global aggregation to update its local model using all shared models within the verification block.

## *10.2.2 Cluster-Based Federated Learning*

### 10.2.2.1 Neural Network Architecture

Extracting informative features from sensor data is crucial for developing effective blade icing detection models. Model design must balance model size, due to bandwidth constraints in gradient exchange, with the capacity to learn relevant features. In this work, feature extraction is achieved using a convolutional neural network (CNN). The structure of the local model, illustrated in Fig. 10.2, incorporates three stacked CNN blocks. Each block comprises a convolution layer (Conv1D), an attention layer, a batch normalization layer (BN), and a ReLU activation layer.

For an input $X_{raw} \in \mathbb{R}^{m \times T}$, where $m$ is the input dimension and $T$ is the sample window size, the output of each CNN block component can be described as

$$\begin{aligned} X_c &= Conv1D(X_{raw}) \\ X_{attn} &= Attention(X_c) \\ X_{out} &= ReLU(BN(X_{attn})), \end{aligned} \tag{10.1}$$

where $X_c$, $X_{attn}$, and $X_{out}$ represent the outputs of the Conv1D, Attention, BN, and ReLU layers, respectively. These variables have a shape of $\mathbb{R}^{F \times T}$, where $F$ is the number of filters in the Conv1D layer.

The attention layer, detailed in Fig. 10.3, operates on the output of the Conv1D layer, which is of shape $\mathbb{R}^{F \times T}$. Here, $F$ represents the number of filters or channels. In a standard Conv1D layer, all channels are treated equally, potentially leading to loss of important information [15, 16]. Thus, an attention mechanism is designed to selectively emphasize informative channels.

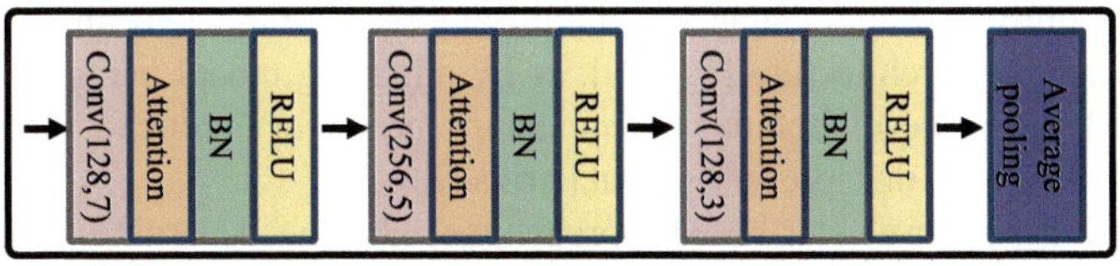


**Fig. 10.2** Details of neural network in deep learning model

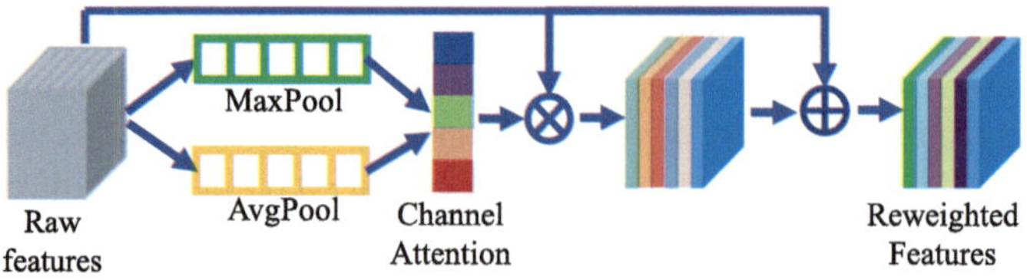


**Fig. 10.3** Illustration of channel calibration attention module

Let the output of the Conv1D layer be $X_c = [x_1, x_2, ..., x_F]$, with each channel $x_i \in \mathbb{R}^{T\times 1}$. To identify significant channels, global average and max pooling are applied to derive distinctive features, $X_{ap}$ and $X_{mp}$, respectively (Fig. 10.3). These pooled features are then processed through a Conv1D layer (not explicitly shown in Fig. 10.3) to compute channel attention. Finally, channel attention is used to refine channel features. The attention layer process is mathematically described as

$$\begin{aligned} X_{mp} &= Conv1D(MaxPool(X_{raw})), \\ X_{ap} &= Conv1D(AvgPool(X_{raw})), \\ \alpha &= \sigma(X_{mp} + X_{ap}), \\ \widetilde{X} &= X_{raw} + \alpha \otimes X_{raw}, \end{aligned} \tag{10.2}$$

where $X_{raw}$ is the attention layer input, $X_{mp}$ and $X_{ap}$ are re-weighted features from average and max pooling, $\alpha$ is the calculated attention weights, $\sigma$ is the sigmoid activation function, $\otimes$ denotes element-wise multiplication, and $\widetilde{X}$ represents the weighted features.

#### 10.2.2.2 Cluster-Based Imbalanced Learning

Data imbalance in training datasets can degrade model performance, especially for minority classes. Common approaches to address imbalance operate at two levels: data and algorithm [17]. Data-level methods typically preprocess raw data using techniques like resampling and data augmentation. These methods can balance data by adjusting the number of samples per class [18]. However, data-level approaches are unsuitable for FL as they require additional information, such as data distribution, which can lead to data leakage and privacy violations [17]. Algorithm-level methods, conversely, mitigate data imbalance by modifying the learning algorithm itself, such as neural network architecture [19, 20], or by adjusting model parameters during training. Algorithm-level methods are more appropriate for FL due to minimal or no raw data processing, thus preserving data privacy.

BLADE employs a cluster-based approach for imbalanced learning, aligning with algorithm-level methodologies, as shown in Client 1's training process in Fig. 10.1. Imbalanced raw data leads to imbalanced features extracted by the neural network.

Therefore, we first establish a cluster for each class, which is then used to build a classifier.

Given learned features $\mathcal{H} = \{X^i\}_{i=1}^{C}$, where $C$ is the number of classes and $X^i \in \mathbb{R}^{N_i \times F \times L}$ is the sample collection for class $i$, with $N_i$ samples, the class centroid $\mathbf{c}_i$ is calculated as

$$\mathbf{c}_i = \text{mean}(H^i) = \frac{1}{N_i}\sum^{N_i} H^i, \quad \mathbf{c}_i \in \mathbb{R}^{F \times L}. \tag{10.3}$$

Once centroids are computed for each class, the probability of a given time-series sample belonging to a class can be determined. For a time series with learned feature $H_s \in \mathbb{R}^{F \times L}$, the probability of belonging to class $i$ is calculated using Euclidean distance:

$$p(y = i|H_s) = \frac{\exp(-\text{Euclidean\_dist}(H_s, \mathbf{c}_i)))}{\sum_k \exp(-\text{Euclidean\_dist}(H_s, \mathbf{c}_k)))}, \tag{10.4}$$

where Euclidean_dist is the Euclidean distance function between $H_s$ and $\mathbf{c}_i$.

Each client model is then trained using the categorical cross-entropy loss function:

$$\mathcal{L} = -\sum_{(x_k, y_k)} \log(p(y = y_k|x_k, \mathbf{c})), \tag{10.5}$$

where $(x_k, y_k)$ represents a sample and $\mathbf{c}$ is the centroid for that sample. Local model parameters are optimized by minimizing the average loss using gradient descent with back-propagation. The Adam optimizer [21] is used, with a learning rate and weight decay fixed at 1e-4 for local training.

#### 10.2.2.3 Federated Aggregation

In the FL framework, each client trains a model locally on its dataset without sharing the raw data with a central server. The gradient of the $k$-th local model in the $c$-th communication round is calculated as

$$\nabla \boldsymbol{\varphi}_{k,c} = \frac{\partial \mathcal{L}_k(\boldsymbol{\varphi}_{k,c})}{\partial \boldsymbol{\varphi}_{k,c}}, \tag{10.6}$$

where $\boldsymbol{\varphi}_{k,c}$ is the gradient of the $k$-th local model at round $c$ and $L_k$ is the categorical cross-entropy loss function defined in Eq.(10.5).

All clients send their model gradients to a central aggregator to construct the global model:

$$\nabla \boldsymbol{\varphi}_c = \sum_{k=1}^{N} \alpha_k \nabla \boldsymbol{\varphi}_{k,c}, \tag{10.7}$$

where $N$ is the number of client models and $\alpha_k$ is the weight of the $k$-th local gradient, proportional to the data size $S_k$ of each client: $\alpha_k = S_k / \sum_{i=1}^{N} S_i$, to account for potential data size variations across clients.

### *10.2.3 Security-Enhanced Blockchain*

BLADE focuses on collaborative blade icing detection learning among $\mathcal{K}$ wind farms in different locations. We assume all participants are potentially dishonest, posing several threats: Firstly, compromised model update quality, where biased or inaccurate updates reduce aggregated model utility. Secondly, data privacy risks, with participants attempting to infer or extrapolate sensitive data from model updates. Thirdly, data authority management concerns, where owners lose control once models are shared, potentially leading to unauthorized model distribution. These threats, primarily originating from the local side, motivate the use of blockchain to mitigate malicious client attacks.

#### 10.2.3.1 Validation Mechanism

Conventional blockchain offers limited malicious node identification. Therefore, BLADE incorporates a validation mechanism proposed by Chen et al. [14]. This mechanism uses majority voting to filter out distorted models from potentially malicious or compromised workers during global model construction. Validators assess local updates from associated workers, flagging updates with significant accuracy drops compared to previous rounds as potentially contaminated. Conversely, updates showing improvement or minor decline are considered legitimate. The threshold $T$ for accuracy-drop measurement is a tunable hyperparameter.

#### 10.2.3.2 Consensus Mechanism

The consensus mechanism protects local model updates for legitimate learning, ensuring their secure recording on the blockchain for global model updates. Preventing malicious participants from becoming miners is crucial, as miners aggregate voting results and record them in blocks. BLADE employs two consensus mechanisms: proof of work (PoW) and proof of stake (PoS).

PoW, a classic blockchain mechanism, involves miners competing in hash operations (typically SHA-256) using computational resources. The first miner to find a hash value below a target threshold appends a new block to the blockchain and receives rewards. "Difficulty" in PoW adjusts the computational effort needed to mine new blocks. PoS, developed to address PoW's resource intensity, uses "coinage", calculated from unused assets and time since the last win, to reduce resource consumption during competition [22].

Reward mechanisms for workers, validators, and miners are adapted from [14]: (1) Worker rewards are proportional to training sample size and local epochs, but are forfeited if validation fails. (2) Validator rewards are based on verifying worker signatures and generating signature-verified updates. (3) Miner rewards are calculated by verifying validator signatures. These reward mechanisms apply to both PoW and PoS consensus algorithms.

### *10.2.4 Privacy Analysis*

BLADE protects privacy through several mechanisms. Firstly, permissioned blockchain replaces trusted servers, preventing high-risk centralized data leakage. Secondly, a validation mechanism defends against invalid data/models from dishonest participants, ensuring only qualified models are used. Thirdly, only trained models are uploaded and transferred via the permissioned blockchain, with local data remaining local under owner control. Cryptographic algorithms further enhance data security. Fourthly, the cluster-based imbalanced learning module protects against data leakage by using cluster centers for classification, improving attack defense capabilities.

## 10.3 Experiment

All models were implemented using Pytorch (V1.7.1, CUDA 11.0) and experiments were conducted on a server equipped with a Tesla V100 GPU with 32 GB video memory.

### *10.3.1 Experimental Settings and Evaluation Metrics*

#### 10.3.1.1 Data Description

Data were collected from two wind farms in Shanxi and Henan Provinces, China, approximately 700 km apart, both experiencing blade icing in winter. Five wind turbines were selected from each farm for model evaluation. Sensors within each turbine, sampling at 5-second intervals, monitor various parameters. Expert consultation identified 16 parameters highly relevant to blade icing, listed in Table 10.1. Icing detection was treated as a binary classification problem, with sensor data labeled by experts.

Data collection periods were 360 and 384 hours for the two farms, respectively. For model training and testing, 60% of the data was used for training and 40% for testing. Performance evaluation was conducted on all test data from each turbine.

**Table 10.1** Specification of the used input parameters

| No. | Variable name | Description |
|---|---|---|
| 1 | wind_speed | Wind speed |
| 2 | wind_direction | Wind direction |
| 3 | generator_speed | Generator speed |
| 4 | power | Active power |
| 5 | yaw_position | Yaw position |
| 6 | pitch1_angle | Angle of pitch 1 |
| 7 | pitch2_angle | Angle of pitch 2 |
| 8 | pitch3_angle | Angle of pitch 3 |
| 9 | pitch1_speed | Speed of pitch 1 |
| 10 | pitch2_speed | Speed of pitch 2 |
| 11 | pitch3_speed | Speed of pitch 3 |
| 12 | environment_temp | Environment temperature |
| 13 | internal_temp | Internal temperature of nacelle |
| 14 | pitch1_moto_tmp | Temperature of pitch motor 1 |
| 15 | pitch2_moto_tmp | Temperature of pitch motor 2 |
| 16 | pitch3_moto_tmp | Temperature of pitch motor 3 |

#### 10.3.1.2 Evaluation Metrics

Model performance was quantified using $F_2$, **Precision**, **Recall**, and **Matthews correlation coefficient** (**MCC**).

$$Precision = \frac{TP}{TP + FP}, \quad Recall = \frac{TP}{TP + FN}, \tag{10.8}$$

$$F_2 = \frac{(1 + 2^2) \times Precision \times Recall}{2^2 Precision + Recall}, \tag{10.9}$$

$$MCC = \frac{TP \times TN - FP \times FN}{\sqrt{(TP + FP)(TP + FN)(TN + FP)(TN + FN)}}, \tag{10.10}$$

where $TP$, $FP$, $FN$, and $TN$ are true positives, false positives, false negatives, and true negatives, respectively. $\beta$ in $F_\beta$ was set to 2. These metrics are standard for blade icing detection algorithm evaluation [23], with $F_2$ and MCC particularly suitable for imbalanced classification. Average metric values across all participants in all communication rounds were used for performance comparison to ensure robustness.

#### 10.3.1.3 Implementation Details

The learning rate for local model training was set to 1e-4. Communication rounds and local training epochs were both set to 20. Blockchain implementation was based on [14]. Imbalance ratios of 20:1, 50:1, and 100:1 were used, defined as $\rho = \frac{S_n}{S_i}$, where $S_n$ and $S_i$ are normal and icing sample counts, respectively. All test sets had a 10:1 imbalance ratio. Experiments were repeated three times.

### 10.3.2 Convergence Analysis

This section examines model convergence across three imbalance ratios, as shown in Fig. 10.4. No malicious clients were present, and worker, validator, and miner counts were randomly selected each communication round. The validation threshold was set to 1. Figure 10.4 shows that the model achieves acceptable performance as communication rounds increase, with greater fluctuations at $\rho = 50 : 1$ and $\rho = 100 : 1$. Performance improvement plateaus after approximately 15 communication rounds, justifying the choice of 20 rounds, considering communication costs in remote wind farm locations.

Figure 10.4 also reveals a strong correlation between accuracy and imbalance ratio; higher ratios correlate with decreased global model accuracy. Greater imbalance reduces icing information available for learning, limiting model performance. Notably, performance is similar at imbalance ratios of 50:1 and 100:1.

### 10.3.3 Effectiveness of Blockchain for Federated Learning

To demonstrate blockchain's effectiveness in federated learning, this section compares model performance under attacks, PoW versus PoS strategies, and the validation mechanism's efficacy.

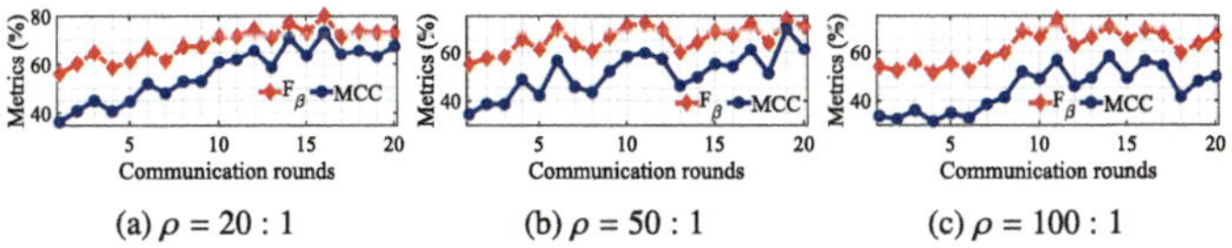


**Fig. 10.4** Convergence analysis in different imbalance ratios

#### 10.3.3.1 Performance Against Poisoning Attacks

BLADE was compared against three models: (1) **FedAvg**: a classic FL model using all client gradients for global model generation, with the same client model as BLADE [11]. (2) **BLADE_IM**: BLADE without the imbalanced learning module. (3) **FedAvg_IM**: FedAvg without the imbalanced learning module. Malicious device ratios were set to $\{20\%, 40\%, 60\%, 80\%\}$, with other settings consistent with previous sections. Attacks involved adding Gaussian noise (variance 2) to client models, and final aggregated model performance was compared.

Results in Fig. 10.5 indicate that BLADE only marginally outperforms FedAvg. Similarly, BLADE_IM is only slightly better than FedAvg_IM. This marginal improvement is attributed to blockchain. However, comparing BLADE to BLADE_IM, and FedAvg to FedAvg_IM, significant performance decreases are observed, indicating the imbalanced learning module's effectiveness in handling both attacks and imbalance. The cluster-based approach uses cluster centers for classification, which are less affected by gradient-level attacks. Performance decreases with increasing malicious client counts, but with only 10 clients, the impact of malicious clients warrants further investigation.

#### 10.3.3.2 PoW Versus PoS Performance

This section compares PoW and PoS mechanisms in BLADE. PoW difficulty was set to 2, and malicious client ratio to 50%, following [14]. Worker, validator, and miner counts were randomized each round, and the validation threshold was 1. Results in Table 10.2 show PoS outperforming PoW at imbalance ratios of 50:1 and 100:1.

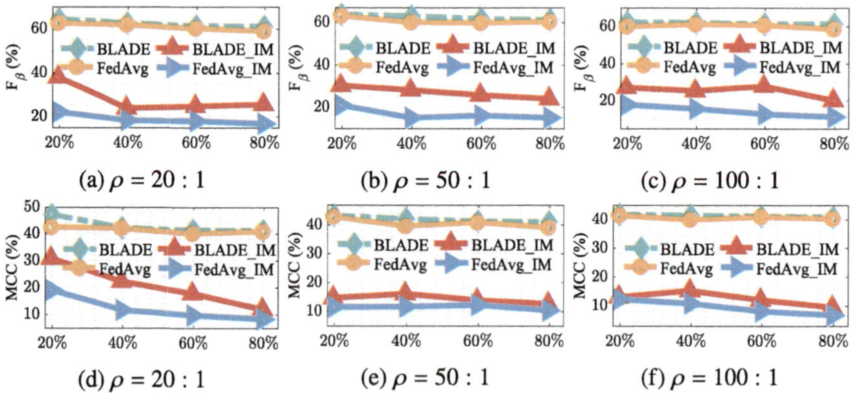


**Fig. 10.5** Performance comparison of different malicious devices in the three imbalance ratios

**Table 10.2** Performance comparison of PoW and PoS

| Metrics | $\rho = 20:1$ | | $\rho = 50:1$ | | $\rho = 100:1$ | |
|---|---|---|---|---|---|---|
| | PoW | PoS | PoW | PoS | PoW | PoS |
| $F_2$ | **66.8** | 64.5 | 60.6 | **62.4** | 59.4 | **60.6** |
| Precision | **34.4** | 29.2 | 26.4 | **27.5** | 24.7 | **26.1** |
| Recall | 87.4 | **92.4** | 90.1 | **92.3** | **92.2** | 90.7 |
| MCC | **48.3** | 44.5 | 40.3 | **42.2** | 38.7 | **40.2** |

**Table 10.3** Performance comparison in different validation threshold (T = validation threshold)

| Metrics | $\rho = 20:1$ | | $\rho = 50:1$ | | $\rho = 100:1$ | |
|---|---|---|---|---|---|---|
| | T=1 | T=10 | T=1 | T=10 | T=1 | T=10 |
| $F_2$ | **62.8** | 60.6 | **61.4** | 59.9 | **60.9** | 59.3 |
| Precision | **27.5** | 26.5 | **27.3** | 25.6 | **26.0** | 24.9 |
| Recall | 88.1 | **90.0** | **90.8** | 90.2 | **92.1** | 91.2 |
| MCC | **42.9** | 40.3 | **41.4** | 39.5 | **40.5** | 38.7 |

#### 10.3.3.3 Validation Mechanism Effectiveness

The validation mechanism's effectiveness was assessed by comparing performance at validation thresholds T=1 and T=10. Worker, validator, and miner counts were set to $\{6, 2, 2\}$, respectively, with a 50% malicious device ratio. Other settings were consistent. Table 10.3 shows that a smaller T (T=1) yields better performance, as expected, across all imbalance ratios, indicating more effective malicious device identification.

### 10.3.4 Comparison with State-of-the-Art Class-Imbalance Methods

To address data imbalance, BLADE was compared to state-of-the-art class-imbalance methods:

(1) **Focalloss**: An algorithm-level loss function [19], with $\alpha = 1$ and $\gamma = 2$. (2) **Class balance**: An improved focal loss using effective sample counts, accounting for data overlap [20]. (3) **LDAM** (label-distribution-aware margin): A margin-based class-balance algorithm enhancing CE loss with prior strategies and a specific training schedule [24]. (4) **WeightedCE**: A CE loss variant weighed by class sample counts for imbalance learning.

These methods replaced BLADE's imbalance module, keeping other components constant. Training settings matched previous sections, with honest clients and a validation threshold of 1. Communication rounds and local epochs were 20. Table 10.4

**Table 10.4** Performance comparison of imbalance learning algorithms in different number of imbalance ratios

| Method | $\rho = 20:1$ | | | | $\rho = 50:1$ | | | | $\rho = 100:1$ | | | |
|---|---|---|---|---|---|---|---|---|---|---|---|---|
| | $F_\beta$ | Precision | Recall | MCC | $F_\beta$ | Precision | Recall | MCC | $F_\beta$ | Precision | Recall | MCC |
| Focalloss | 37.8 | 66.6 | 34.4 | 43.9 | 27.1 | **75.3** | 23.6 | 38.6 | 17.0 | **70.0** | 14.6 | 27.7 |
| Class-balance | 42.5 | 13.4 | **99.6** | 16.7 | 48.6 | 16.4 | **99.1** | 26.3 | 44.7 | 14.3 | **99.3** | 21.2 |
| LDAM | 48.3 | **67.2** | 45.6 | 51.4 | 28.8 | 62.9 | 26.0 | 35.8 | 21.1 | 64.4 | 18.6 | 29.6 |
| WeightedCE | 53.0 | 54.3 | 53.2 | 48.8 | 51.9 | 58.0 | 51.3 | 49.7 | 47.4 | 56.8 | 46.0 | **46.4** |
| *Ours* | **68.6** | 49.0 | 79.1 | **56.2** | **65.3** | 42.6 | 79.6 | **50.8** | **61.7** | 34.2 | 79.7 | 44.2 |

shows BLADE achieving the highest MCC, except at 100:1 where WeightedCE is slightly better. BLADE has the best $F_2$ across all ratios. WeightedCE is second best, suggesting simple methods can be effective, especially when clients have uniform imbalance ratios. Algorithm-level methods (Focalloss, Class balance, LDAM) underperformed, possibly due to their design for central, not distributed, learning, leading to performance degradation in FL.

## *10.3.5 Impacts of Different Settings in the Neural Network*

### 10.3.5.1 Impacts of Neural Network Structure

Model learning capability is crucial for BLADE. Five baseline models were used to assess local model structure impacts: (1) **MLP (multilayer perceptron)**: Three FC layers with 500 hidden units each, and dropout layers. (2) **LSTM** (Long Short-Term Memory): Single-layer LSTM with 8 hidden units. (3) **CNN**: Single-layer CNN with 128 filters. (4) **DenseNet**: DenseNet structure without attention modules [15]. (5) **MLSTM-FCN**: Settings from the original paper [25]. These replaced BLADE's neural network with other settings as in Sect. 10.3.2.

Table 10.5 shows BLADE performing best in $F_2$ except at $\rho = 100 : 1$ (second to Densenet). In MCC, BLADE is also competitive, except at $\rho = 100 : 1$ (third, behind DenseNet and LSTM). MLP and LSTM performance is similar across ratios. Increased model complexity (DenseNet, MLSTM_FCN) did not consistently improve performance, indicating that suitable model design is more critical than complexity.

### 10.3.5.2 Comparison with Other Attention Mechanisms

To evaluate the proposed attention module, it was compared with four others (Table 10.6): **SE** (Squeeze-and-Excitation) [26], **GC** (Global Context) [27], **TR** (Temporal Relational) [28], and **CBAM** (Convolutional Block Attention Module) [29]. These replaced BLADE's attention module, with other settings consistent with Sect. 10.3.2.

Table 10.6 shows BLADE performing best at $\rho = 20 : 1$. At $\rho = 50 : 1$ and $\rho = 100 : 1$, BLADE is not consistently best, except for MCC at $\rho = 50 : 1$. GC performs best at $\rho = 100 : 1$, with BLADE second.

## *10.3.6 Ablation and Sensitivity Analysis*

Ablation analysis assessed component importance using two variants: (1) **No_IM**: without the imbalance learning module. (2) **No_Attn**: without the channel attention module. Settings matched Sect. 10.3.2. Figure 10.6 shows the largest performance

**Table 10.5** Performance comparison of local model

| Method | $\rho = 20:1$ | | | | $\rho = 50:1$ | | | | $\rho = 100:1$ | | | |
|---|---|---|---|---|---|---|---|---|---|---|---|---|
| | $F_\beta$ | Precision | Recall | MCC | $F_\beta$ | Precision | Recall | MCC | $F_\beta$ | Precision | Recall | MCC |
| MLP | 61.2 | 42.1 | 70.9 | 48 | 58.8 | 39.2 | 68.9 | 45.0 | 57.1 | 37.6 | 68.1 | 42.9 |
| LSTM | 61.5 | 48.8 | 68.0 | 51.5 | 58.5 | **45.4** | 66.1 | 47.8 | 59.2 | **40.4** | 71.0 | 45.9 |
| CNN | 43.9 | 17.1 | 72.5 | 21.9 | 45.8 | 17.6 | 76.9 | 23.8 | 44.1 | 16.4 | 76.7 | 21.9 |
| DenseNet | 66.1 | 34.7 | **88.0** | 47.9 | 64.2 | 32.6 | **88.6** | 45.7 | **66.9** | 35.0 | **90.0** | **48.8** |
| MLSTM_FCN | 64.1 | 45.0 | 76.1 | 51.1 | 63.3 | 34.6 | 83.7 | 45.6 | 59.2 | 34.8 | 78.0 | 42.9 |
| *Ours* | **68.6** | **49.0** | 79.1 | **56.2** | **65.3** | 42.6 | 79.6 | **50.8** | 61.7 | 34.2 | 79.7 | 44.2 |

**Table 10.6** Performance comparison of attention modules

| Method | $\rho = 20:1$ | | | | $\rho = 50:1$ | | | | $\rho = 100:1$ | | | |
|---|---|---|---|---|---|---|---|---|---|---|---|---|
| | $F_\beta$ | Precision | Recall | MCC | $F_\beta$ | Precision | Recall | MCC | $F_\beta$ | Precision | Recall | MCC |
| SE | 67.1 | 50.5 | 76.5 | 55.8 | **65.7** | 38.5 | **83.2** | 49.1 | 60 | 31.9 | 80.9 | 42.1 |
| GC | 64.7 | 42.5 | 78.3 | 50.2 | 62.8 | 38 | 79.3 | 46.8 | **64.1** | **38.7** | **81.3** | **48.3** |
| TR | 53.5 | **51.0** | 55.8 | 47.4 | 57.5 | 36.2 | 70.8 | 42.5 | 49.3 | 38.4 | 58.6 | 38.5 |
| CBAM | 57.1 | 36.0 | 75.5 | 41.5 | 55 | 33.4 | 75.4 | 38.4 | 55.4 | 30.8 | 77.2 | 37.8 |
| *Ours* | **68.6** | 49.0 | **79.1** | **56.2** | 65.3 | **42.6** | 79.6 | **50.8** | 61.7 | 34.2 | 79.7 | 44.2 |

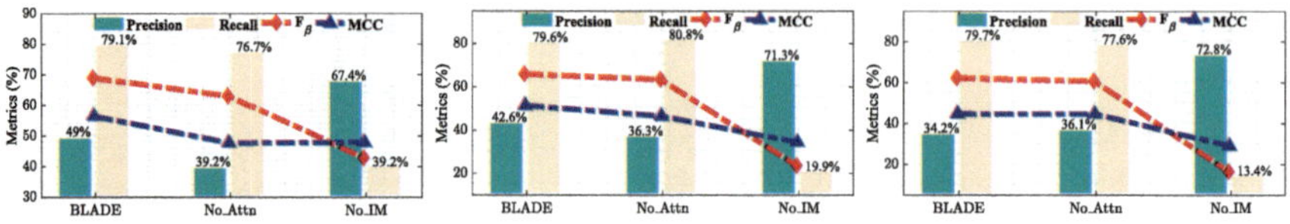


(a) $\rho = 20:1$ (b) $\rho = 50:1$ (c) $\rho = 100:1$

**Fig. 10.6** Ablation analysis

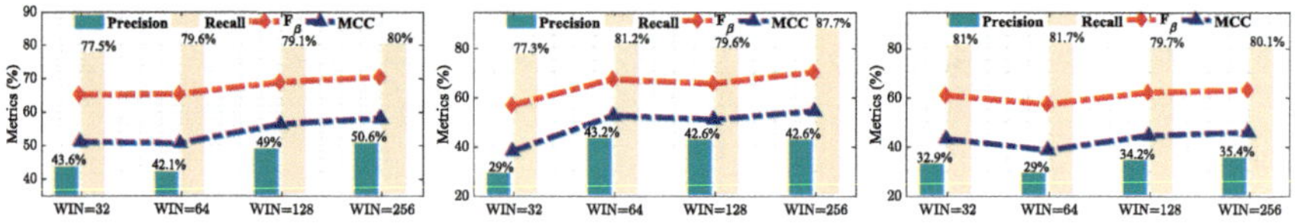


(a) $\rho = 20:1$ (b) $\rho = 50:1$ (c) $\rho = 100:1$

**Fig. 10.7** Sensitivity analysis

drop when the imbalance learning module is removed across all ratios. Removing the attention module caused a smaller decrease. The imbalance learning module is more impactful than the attention module.

Sensitivity analysis evaluated window size influence on accuracy (Fig. 7.3). Window size 256 yielded the best performance across all ratios, likely due to larger windows containing more information (Fig. 10.7).

## 10.4 Conclusion

Wind energy is crucial for sustainable development. Blade icing significantly limits wind turbine performance, especially in winter. This paper explored BLADE, a blockchain-integrated imbalanced federated learning model for wind turbine blade icing detection, addressing centralized server limitations in conventional FL. BLADE incorporates a novel cluster-based imbalance learning module to handle data imbalance. A blockchain validation mechanism enhances defense against poisoning attacks. BLADE's performance was comprehensively evaluated against state-of-the-art class-imbalanced learning methods, and different neural networks and attention mechanisms. Blockchain effectiveness and BLADE component contributions were also investigated.

However, participant heterogeneity was not considered, potentially reducing model performance. Future work will design a heterogeneous blockchain-based federated learning model and explore icing severity detection, potentially through output probability interpretation or severity labeling based on domain knowledge.

## References


1. Ren Z, Verma AS, Li Y, Teuwen JJ, Jiang Z (2021) Offshore wind turbine operations and maintenance: a state-of-the-art review. Renew Sustain Energy Rev 144:110886
2. Wei K, Yang Y, Zuo H, Zhong D (2020) A review on ice detection technology and ice elimination technology for wind turbine. WindÂ Energy 23
3. Dalili N, Edrisy A, Carriveau R (2009) A review of surface engineering issues critical to wind turbine performance. Renew Sustain Energy Rev 13(2):428–438
4. Battisti L, Baggio P, Fedrizzi R (2006) Warm-air intermittent de-icing system for wind turbines. Wind Eng 30(5):361–374
5. Shajiee S, Pao LY, McLeod RR (2014) Monitoring ice accumulation and active de-icing control of wind turbine blades. In: Wind turbine control and monitoring. Springer, pp 193–230
6. Wang P, Zhou W, Bao Y, Li H (2018) Ice monitoring of a full-scale wind turbine blade using ultrasonic guided waves under varying temperature conditions. Struct Control Health Monitoring 25(4):e2138
7. Roberge P, Lemay J, Ruel J, Bégin-Drolet A (2018) A new atmospheric icing detector based on thermally heated cylindrical probes for wind turbine applications. Cold Regions Sci Technol 148:131–141
8. Corradini ML, Cristofaro A, Pettinari S (2016) "A model-based robust icing detection and estimation scheme for wind turbines. In: European control conference (ECC). IEEE, pp 1451–1456
9. Cheng X, Shi F, Liu X, Zhao M, Chen S (2021) A novel deep class-imbalanced semisupervised model for wind turbine blade icing detection. IEEE Trans Neural Netw Learn Syst
10. Goncalves C, Bessa RJ, Pinson P (2021) Privacy-preserving distributed learning for renewable energy forecasting. IEEE Trans Sustain Energy
11. McMahan B, Moore E, Ramage D, Hampson S, y Arcas BA (2017) Communication-efficient learning of deep networks from decentralized data. In: Artificial intelligence and statistics. PMLR, pp 1273–1282
12. Medhane DV, Sangaiah AK, Hossain MS, Muhammad G, Wang J (2020) Blockchain-enabled distributed security framework for next-generation iot: an edge cloud and software-defined network-integrated approach. IEEE Internet Things J 7(7):6143–6149
13. Wu Y, Dai H-N, Wang H (2020) Convergence of blockchain and edge computing for secure and scalable iiot critical infrastructures in industry 4.0. IEEE Internet Things J 8(4):2300–2317
14. Chen H, Asif SA, Park J, Shen C-C, Bennis M (2021) Robust blockchained federated learning with model validation and proof-of-stake inspired consensus. arXiv preprint arXiv:2101.03300
15. Cheng X, Li G, Ellefsen AL, Chen S, Hildre HP, Zhang H (2020) A novel densely connected convolutional neural network for sea-state estimation using ship motion data. IEEE Trans Instrumen Measur 69(9):5984–5993
16. Zhang J, Chen J, Cao J, Liu R, Bian L, Chen S (2021) Dual attention granularity network for vehicle re-identification. Neural Comput Appl, 1–12
17. Wang L, Xu S, Wang X, Zhu Q (2020) Towards class imbalance in federated learning. arXiv preprint arXiv:2008.06217
18. Xie Y, Qiu M, Zhang H, Peng L, Chen Z (2020) Gaussian distribution based oversampling for imbalanced data classification. IEEE Trans Knowl Data Eng
19. Lin T-Y, Goyal P, Girshick R, He K, Dollár P (2017) Focal loss for dense object detection. In: Proceedings of the IEEE international conference on computer vision, pp 2980–2988
20. Cui Y, Jia M, Lin T-Y, Song Y, Belongie S (2019) Class-balanced loss based on effective number of samples. In: Proceedings of the IEEE conference on computer vision and pattern recognition, pp 9268–9277
21. Kingma DP, Ba J (2014) Adam: a method for stochastic optimization. arXiv preprint arXiv:1412.6980
22. Cao B, Zhang Z, Feng D, Zhang S, Zhang L, Peng M, Li Y (2020) Performance analysis and comparison of pow, pos and dag based blockchains. Digital Commun Networks 6(4):480–485

23. Cheng X, Shi F, Zhao M, Li G, Zhang H, Chen S (2021) Temporal attention convolutional neural network for estimation of icing probability on wind turbine blades. IEEE Trans Indus Electron
24. Cao K, Wei C, Gaidon A, Arechiga N, Ma T (2019) Learning imbalanced datasets with label-distribution-aware margin loss. In: Advances in neural information processing systems, pp 1567–1578
25. Karim F, Majumdar S, Darabi H, Harford S (2019) Multivariate lstm-fcns for time series classification. Neural Networks 116:237–245
26. Hu J, Shen L, Sun G (2018) Squeeze-and-excitation networks. In: Proceedings of the IEEE conference on computer vision and pattern recognition, pp 7132–7141
27. Cao Y, Xu J, Lin S, Wei F, Hu H (2019) "Gcnet:non-local networks meet squeeze-excitation networks and beyond. In: IEEE/CVF international conference on computer vision workshop (ICCVW), pp 1971–1980
28. Hao Y, Cao H (2020) A new attention mechanism to classify multivariate time series. In: IJCAI, pp 1999–2005
29. Woo S, Park J, Lee J-Y, So Kweon I (2018) Cbam: convolutional block attention module. In: Proceedings of the European conference on computer vision (ECCV), pp 3–19

# Chapter 11
# Conclusion Remarks

This book, "Computational Methods for Blade Icing Detection of Wind Turbines", embarked on a critical exploration of advanced techniques to address a challenge of paramount importance to the sustainable energy sector: blade icing. As wind power generation continues its exponential growth, particularly in colder and high-altitude regions, the detrimental effects of ice accumulation on turbine blades become increasingly pronounced. From reduced energy output and increased maintenance costs to heightened safety risks and potential structural damage, blade icing stands as a significant impediment to the reliable and economically viable operation of wind farms.

Our journey through the preceding chapters has demonstrated the transformative potential of computational methodologies in tackling this complex problem. We have moved beyond traditional, often resource-intensive and less adaptable, approaches to embrace the cutting-edge capabilities of data-driven techniques, particularly those rooted in deep learning, graph neural networks, and federated learning.

Part I laid the groundwork by delving into the realm of **deep learning-based methods**. We showcased the efficacy of Temporal Attention Convolutional Neural Networks (TACNN) in intelligently discerning sensor and time-step significance, enabling more precise icing detection. The exploration of Multilevel Convolutional Recurrent Neural Networks (MCRNN) highlighted the value of multi-scale feature extraction and the robust handling of imbalanced data, a pervasive challenge in real-world icing datasets. Furthermore, the introduction of a Novel Deep Class-Imbalanced Semi-supervised Model (DCISS) offered a compelling solution for leveraging both labeled and unlabeled data, maximizing learning efficiency in scenarios where labeled data is scarce. This section underscored the power of deep learning architectures to automatically extract complex features, capture temporal dynamics, and navigate the intricacies of imbalanced datasets, paving the way for more accurate and efficient icing detection systems.

 
X. Cheng et al., *Computational Methods for Blade Icing Detection of Wind Turbines*, Engineering Applications of Computational Methods 24,
https://doi.org/10.1007/978-981-96-6763-5_11

Part II transitioned to **graph neural network-based methods**, recognizing the inherent interconnectedness of wind turbine systems. We presented the Attention-Enhanced Graph Convolutional Neural Network (GTAN), showcasing its ability to adaptively analyze sensor correlations and enhance feature learning for superior icing recognition. The exploration of Multilevel Graph Neural Networks (MWGCN) expanded on this foundation, demonstrating the synergistic combination of wavelet analysis and graph neural networks to improve feature learning across multiple scales, further enhanced by a scale-attention mechanism. These chapters illuminated the potential of graph learning in modeling complex relationships within wind turbine sensor data, offering a nuanced understanding of spatial-temporal patterns crucial for accurate icing detection.

Finally, Part III ventured into the promising domain of **federated learning-based methods**. Addressing the critical need for data privacy and collaborative model training across geographically dispersed wind farms, we presented the graph temporal attention network for imbalanced wind turbine blade icing prediction, leveraging graph attention networks and temporal attention mechanisms for enhanced performance. The introduction of federated learning for blade icing detection explored heterogeneous federated learning models (BiFL), designed to function effectively across diverse wind turbine types while addressing data imbalance. Furthermore, the exploration of FedBIP: A Federated Learning-Based Model for Wind Turbine Blade Icing Prediction showcased feature selection and knowledge distillation techniques to further enhance training and prediction performance. Culminating in the investigation of blockchain-empowered clustered federated Learning, we demonstrated how blockchain technology could bolster the security and robustness of federated learning for blade icing detection, while employing novel imbalanced learning techniques to enhance detection accuracy. This section highlighted the transformative potential of federated learning in enabling collaborative, privacy-preserving model training, paving the way for scalable and robust icing detection systems across distributed wind farm networks.

In summary, this book has presented a comprehensive and multifaceted exploration of computational methods for blade icing detection. We have demonstrated the efficacy of advanced techniques in addressing the inherent complexities of this challenge, offering practical solutions and innovative methodologies. Through rigorous experimental evaluations using both publicly available benchmark datasets and real-world SCADA data from operational wind turbines, we have validated the superior performance of the proposed methods compared to existing state-of-the-art approaches. Our findings underscore the potential of these advanced computational strategies to accurately forecast blade icing occurrences, contributing to reduced maintenance costs, enhanced operational efficiency, and improved safety within wind farms.

Looking ahead, several promising avenues for future research emerge from this work. Further exploration of **hybrid approaches**, combining the strengths of deep learning, graph neural networks, and federated learning, could lead to even more robust and adaptable icing detection systems. Investigating methods to **predict the**

**severity of icing**, rather than simply detecting its presence, remains a crucial direction for future research, enabling more nuanced and proactive mitigation strategies. Exploring the **generalizability of these models across diverse wind turbine types and geographical locations** is also paramount to ensure widespread applicability and impact. Furthermore, the integration of **explainable AI (XAI) techniques** to enhance the interpretability of these complex models could build trust and facilitate wider adoption by wind farm operators. Finally, continued research into **optimizing computational efficiency and reducing communication overhead** in federated learning frameworks will be essential to ensure practical deployment in resource-constrained environments.

In conclusion, the advanced computational methodologies presented in this book represent a significant step forward in the field of blade icing detection. By embracing the power of deep learning, graph neural networks, and federated learning, we have demonstrated the potential to overcome the limitations of traditional approaches and pave the way for more reliable, efficient, and cost-effective solutions. It is our hope that this book serves as a valuable resource and a catalyst for continued innovation in this critical area, ultimately contributing to a more sustainable and resilient wind energy future.

GPSR Compliance

*The European Union's (EU) General Product Safety Regulation (GPSR) is a set of rules that requires consumer products to be safe and our obligations to ensure this.*

*If you have any concerns about our products, you can contact us on ProductSafety@springernature.com*

In case Publisher is established outside the EU, the EU authorized representative is:

Springer Nature Customer Service Center GmbH
Europaplatz 3
69115 Heidelberg, Germany

**Batch number: 10704307**

Printed by Printforce, the Netherlands